TECHNICAL COMMUNICATION
The Practical Craft

THIRD EDITION

MARIS ROZE
DeVRY Institutes

Prentice Hall
Upper Saddle River, New Jersey Columbus, Ohio

Library of Congress Cataloging-in-Publication Data
Roze, Maris.
Technical communication : the practical craft/Maris Roze.—3rd ed.
p. cm.
Includes bibliographical references and index.
ISBN 0-13-455874-X (alk. paper)
1. Communication of technical information. 2. Technical writing. I. Title.
T10.5.R65 1997
8088.0666—dc20 96-5164
CIP

Cover photo: ©guildhaus
Editor: Stephen Helba
Production Editor: Alexandrina Benedicto Wolf
Design Coordinator: Julia Zonneveld Van Hook
Text Designer: Ed Horcharik
Cover Designer: Russ Maselli
Production Manager: Laura Messerly
Marketing Manager: Frank Mortimer, Jr.
Electronic Text Management: Marilyn Wilson Phelps, Matthew Williams, Karen L. Bretz, Tracey Ward

This book was set in Garamond by Prentice Hall and was printed and bound by R.R. Donnelley & Sons Company. The cover was printed by Phoenix Color Corp.

Simon & Schuster/A Viacom Company
Upper Saddle River, New Jersey 07458

Printed in the United States of America

10 9 8 7 6 5 4 3

ISBN: 0-13-455874-X

Prentice-Hall International (UK) Limited, *London*
Prentice-Hall of Australia Pty. Limited, *Sydney*
Prentice-Hall of Canada, Inc., *Toronto*
Prentice-Hall Hispanoamericana, S. A., *Mexico*
Prentice-Hall of India Private Limited, *New Delhi*
Prentice-Hall of Japan, Inc., *Tokyo*
Simon & Schuster Asia Pte. Ltd., *Singapore*
Editora Prentice-Hall do Brasil, Ltda., *Rio de Janeiro*

For students and other explorers of the real world

Preface

The scope and modular design of *Technical Communication: The Practical Craft, Third Edition,* offer instructors several options for using it:

1. The range of types of writing assignments makes the book suitable for a technical writing or a business writing course at the collegiate level.
2. The modular approach to topics allows the instructor to structure the course to emphasize special needs and to adjust the structure in mid-course if early results indicate other needs.
3. Those who teach technical or business subjects, rather than writing, can use the book as a supplement to increase writing in their courses. The modules are organized for self-reference to guide such assignments.

Technical Communication: The Practical Craft is also designed to support effective *teaching* of communication skills.

Much experience has shown that students take writing seriously when they see it as an essential part of their career preparation. They make the best gains when they can write about subjects in their major field and when they perceive assignments as realistic simulations of career requirements rather than academic exercises—hence, the persistent applications orientation of this book.

Theory is included in limited doses so that students can readily apply it to real-world assignments. To a greater extent, the modules focus on writing situations and on the logic of particular formats for those situations. General principles are derived from the need to solve particular problems.

For most students, this inductive approach provides a comfortable fit. In addition to its motivational advantages, the method gives students a chance to learn by doing. Theory becomes a more functional instrument when it is derived from specific cases.

On Evaluation

To help students improve their communication skills, feedback should play an important role. Unfortunately, this element often proves to be a barrier rather than a means to improvement. Assignments may not be made because they are seen as requiring complex evaluative responses. Technical and business instructors may stint on feedback because of uncertainty about "proper" terminology. Evaluations by writing teachers may be perceived as subjective and arbitrary by students.

These pitfalls can be avoided. In the area of feedback, the book's applied focus provides a major advantage. The structured nature of the assignments, as well as the general career orientation, establish a logical framework for evaluation that is helpful to students and sparing of instructors.

General and specific guidelines for evaluation are given in the instructor's manual. The manual also includes suggestions for using the book in writing courses and across the curriculum generally.

To the Student

This is a book dictated by experience. It describes what you need to do in specific writing situations rather than in general cases. The experience on which the book is based is drawn from thousands of young men and women who went to college to prepare for technical and business careers and found that effective writing—and other communication tools—were essential to their education and career plans.

Career education programs are often designed in consultation with business and industry leaders. Increasingly in recent years, these consultants have stressed the importance of clear communication as a basic tool for getting the work done. As the work environment has become more complex and competitive, communication skills have become as important as technical and business skills.

Experience has also shown that communication must be practiced to be learned and that improvement is greatest when communication is applied to meaningful subjects.

The subjects this book points to are the work activities of technical and business careers. The suggested writing topics establish a direction and a basic purpose for your writing. Beyond these suggestions, however, the book structures the writing task into a series of choices leading to effective formats for presentation. Such formats are widely used in the working world because they answer the need for efficient communication. They organize a message so that it is clear and complete to a busy reader. And they help writers achieve their purposes.

You may use this book in a writing class, but you can also use it as a guide for yourself. Most units in Sections 1 to 4 assemble all the information you need to carry out a common writing assignment. The assignment may come from an instructor, but it may also come from yourself when you feel the need. You may wish to write a letter of inquiry, for example, or a set of instructions, or a proposal.

The unit you consult will tell you how to plan and execute the assignment and then show you an example or two. For help with the general rules of writing and questions about form and procedure, refer to Section 6.

Two sections offer units on topics of special importance. Section 4 deals with oral communication and includes units that discuss active listening as well as speaking. Section 5 focuses on research as a stage of the development cycle that may include outlines and bibliographies and that applies useful techniques such as electronic searching and word processing.

Most units also include a paragraph or two on the benefits of the exercise. They stress the value of this form of communication to people in industry and the advantages to you in meeting their needs. These comments underscore a central premise of this book: Writing—as well as speaking—is a process of thinking about others and doing your best to help them understand your message. The discipline this requires makes writing hard work.

But, as the book tries to show, the effort to communicate brings the greatest rewards to you. The more you write, the better you write. The more you struggle to explain, the better you understand the subject yourself. When you work to improve your writing, you enrich yourself.

With a Little Help from Our Friends

This book really began with our students and what they wanted from us when we asked them to write and speak and listen as though their careers depended on it. Thanks to them for being serious about their education and for challenging us to help make it work.

The revision of this book has been an evolutionary process of several years' duration. For their feedback on the first and second editions and for specific and general suggestions for change, thanks to the dedicated writing instructors of the DeVRY Institutes system. Particular thanks go to the faculty serving in DeVRY's systemwide writing assessment program (WRAP) for their real-world perspectives on evaluation and to Professors Nancy Stegall and Lory Hawkes for leading the way toward more significant use of the Internet and the World Wide Web in our writing programs.

For administrative support and guidance in projects associated with the text, thanks to Marilynn Cason, George Dean, Norm Levine, and Patrick Mayers. A special thanks to my colleagues Amin Karim and Charles Koop for their insights into technology and its role in technical communication. To my son, Robert Roze, I owe gratitude for many things, including a set of user-friendly explanations of the principles of electronic communication.

I also appreciate the helpful comments and suggestions from the following reviewers: Linda Marquardt, Lake Area Technical Institute; Susan H. Chin, DeVry Institute of Technology, Georgia; and Laura Renick-Butera, DeVry Institute, New Jersey.

Maris Roze

Contents

TECHNICAL COMMUNICATION
The Practical Craft

INTRODUCTION

The Development Process

Clear, effective communication about our activities—past and present and planned—has always been a requirement for getting the work done. Just as often, this requirement has been neglected in the press of affairs, and poor communication has held our work back and taken its toll in lost time and opportunities.

Every generation rediscovers the importance of clear communication as it struggles to solve its own problems and to improve the conditions of life. The discovery it makes is that all problems, even technical ones, are ultimately *human* problems because they must be solved through the cooperative efforts of people with differing knowledge, priorities, and points of view.

For technical and business specialists and other professionals, the discovery that communication is the vital link in getting the job done often comes as a surprise. The focus of training for many specialists has been so strongly on technical matters that the needs of communication and coordination are often neglected. But technical achievements themselves have increased the importance of technical communication.

Because of advances in technology and growth in global competition, business and industry have become more dynamic and complex. There is greater specialization as companies look for more efficient ways to serve customers. Quality is a leading priority. Project teams that cut across normal functional lines are increasingly used. Products that are more sophisticated technically must also be more user friendly. All these factors contribute to the need for effective communication between specialists from different fields; just as important is the communication between specialists and nontechnical people, such as managers or customers.

Project teams charged with new product development often combine not only programmers, engineers, technicians, managers, and marketers, but also technical writers and sometimes even users. The writers often influence the design process by focusing attention on user needs through their documentation of the product. In these cases, both writing and product development become collaborative activities.

When maintenance and servicing are performed at customer sites, the level of communication also increases. An understanding of technical problems must be gained from phone or fax information, e-mail, and face-to-face discussion, as well as from direct examination. To many field service specialists, the critical aspect of the job is developing effective communication with the client, who is often under pressure and upset. Field service workers speak of this challenge as "fixing the customer" before turning their attention to the equipment.

Service calls also require follow-up communications. Problems and their solutions must be described in written reports for supervisors and for customers so that a permanent record of service activities is established. Such reports must be extremely precise, since business obligations and even legal issues may be involved.

The Electronic Revolution

Advances in communications and information retrieval technology have also increased the importance of communication. Today, more efficient forms of communication and more productive research tools are offered by such innovations as the Internet, computerized data bases, and CD-ROM resources. These advances increase the stakes as well as the capabilities for communication.

The capabilities of global and virtually instant communication via the Internet are bringing more people into immediate contact on issues and projects, both as initiators of and respondents to messages. The instantaneous, or real-time, nature of the communication has also contributed to the expansion of collaborative and interactive activities, as seen in committees writing joint reports or students reviewing or editing one another's work on their computer screens.

Of equal significance is the Internet's contribution to research capabilities. Search software such as Gopher, Archie, WAIS (Wide Area Information Service), and other tools, along with retrieval capability, can now bring a global wealth of materials to anyone with a computer, a modem, and access to the Internet. These services, in effect, offer a "global digital library"[1] that can provide text, sound, graphics, video, and multimedia combinations of information from corporations, universities, government agencies, think tanks, special interest groups, and individuals world wide.

At the same time, the scope and timeliness of information available from on-line computerized data bases and CD-ROM technologies has vastly increased the efficiency of the research process. These resources not only store large amounts of information but also organize and continually update materials for ready access by simple search strategies.

A Communications Responsibility

Today, the stereotype of the technical specialist working in isolation, scribbling numbers, sketching symbols, and mumbling acronyms is dead. The specialist must be a

1. A term suggested by Douglas E. Comer, *The Internet Book* (Prentice-Hall: Englewood Cliffs, NJ, 1994), 265.

communicator precisely because the work has become more complex, resources more available, and the stakes higher.

Specialists are problem solvers who apply their knowledge, skill, and judgment to new projects or existing operations that must be explained to others, coordinated with others, and approved by others. The writing and speaking and listening this requires are as much a part of the problem to be solved as the technical analysis or design.

The acceptance of communication as a major responsibility of the specialist's job is also a first step toward improving communication skills. But to make real progress in communicating effectively, philosophical acceptance must be followed by practical strategy, and then by lots of application and practice.

A planning-and-development strategy can be applied to most of the formal communications we have to make. Such a strategy is based on a clear understanding of the communication process and on logical analysis of communication tasks. The strategy begins with analysis of three closely related elements of any communication—*subject, purpose,* and *audience.*

DEMYSTIFYING THE TASK

Start with the question of subject. As a technical or business writer, you have a major advantage over the generalist. Your expertise makes it easier to follow the classic advice given to young writers: Write about what you know.

Subject

While you are in college, you are still learning the discipline that will become the basis of your subject when you communicate. In the early terms, one of your assignments might be to write or speak about what you are learning. In later terms, you might be asked to explain not only what you know about the subject but to analyze or evaluate it.

When you are working in business or industry, the subject of your written or spoken message will remain a given in most cases. You will be expected to report on the day-to-day work you do, and you will get special reporting assignments for the special projects you do.

An additional communications element in your career arises from the opportunities or needs you recognize yourself. The work you do will inevitably present problems or reveal deficiencies, either in the work itself or the environment around it. As a professional, you will be expected to bring these problems or opportunities to the attention of others. As you gain experience and responsibility, the initiative in such communication will shift more and more from others to you.

At this point, considerations of *purpose* and *audience* may be more important than your *subject.* Let's look at all three elements in more detail.

Subject and Purpose

Even when you are assigned a subject ("Write about X"), you must decide the purpose of your communication. Beyond fulfilling the assignment, what will you do with the subject? Describe X? Analyze X? Justify X? Consider that X might be any of the following:

process	person	rule
mechanism	group	policy
product	organization	procedure
substance	event	relationship
technique	conflict	issue
design	transaction	trend

For any of these very general subjects (and the specific cases they include), your purpose in writing or speaking may be one or more of the following:

1. *Persuasive*—convince, gain approval, win support, justify
2. *Informative*—explain, describe, list, illustrate, notify, announce
3. *Directive*—instruct or order
4. *Analytical*—determine how something works, establish the relationships between parts, find causes and their effects
5. *Evaluative*—judge, rate, compare and contrast, determine feasibility

You can easily see that these purposes could be combined and that they have some natural overlap. You might analyze a technique to evaluate its effectiveness, you might explain to persuade, and so on.

But notice that in such cases as "explain to persuade" your main purpose can be separated from a supporting one. (Can you say which purpose is primary here?) Before you start writing, spell out precisely what you want to accomplish; the statement will help keep your message in sync with your purpose.

In business and technical writing, the purpose may also be built into the format. A proposal, for example, always has a persuasive general purpose. In particular cases, the proposal may be aimed at achieving a very specific result, which can be specifically stated, for example: "To obtain a contract from the school board to equip a computer learning laboratory for grades K-6 at Ralph Waldo Emerson School in Branchville, NJ." In this competitive bidding case, you would also presumably know the specific format in which the proposal has to be written, as well as other requirements and limits.

In many work situations, a general kind of report, such as a progress report or field report, will be required in a specific format established by the company. That format will also be designed to accomplish fairly definite purposes, such as to allow a supervisor to monitor your performance and identify problems needing intervention or correction.

Purpose and Audience

Your purpose in writing or speaking must also be weighed in relation to the audience you will address. Analyzing your audience may not change your purpose, but it will often change your approach and language.

Start with the most basic question: Who is my audience? Sometimes the obvious answer may be only partly correct. My supervisor might read my progress reports but also attach them to her own reports to a supervisor. I may write for my instructor, but he may ask a graduate assistant to screen all reports first. Multiple audiences are, in fact, common for memos, letters, and informal reports of various kinds.

Sometimes I may not be able to specifically identify the audience at all. An inquiry letter or phone call to a business organization is a typical example. At another level, user's manuals and instructions are often written for audiences that are known only in general terms. In a *computer discussion group* on the Internet, I may send and receive messages within a group of people I have never met but who happen to share my interest in the topic of discussion.

What difference does it make who your audience is? The importance of at least two audience characteristics should be self-evident. One is the level of language and technical understanding that your readers or listeners possess. If you explain too little and your language is too technical, your audience won't understand, and your purpose will not be achieved. If you explain too much for a more knowledgeable audience, they may become bored or irritated, again frustrating your purpose.

The other key consideration involves the motives and interests of your audience and the pressures bearing on them. A technical reader may be primarily concerned with the design or efficiency of your proposal. A first-line supervisor may wonder about its effect on worker morale. An executive may question its cost.

The fact that you can identify the technical level and basic motivations of your audience should at least help you decide the level of detail to use and guide your choice of words. It should also help you choose a strategy of presentation that will best respond to audience concerns. (For more specific discussion of audience types, see the section on Technical Style, pages 15–28.)

A Case for Discussion The school board for District 87 in Branchville, NJ, has issued a request for proposals to equip computer learning laboratories for grades 6–8 at five junior high schools. Bids for computers and tutorial software in language and math have been invited. As required by law, bids will be examined in board sessions open to the public.

Because of an eroding tax base, the board has been facing budget shortfalls and has proposed a 1-year delay in a pay raise won by the teachers' union. At the same time, an influx of undocumented aliens has raised the level of deficiencies in basic skills among students. The learning lab concept is, in part, an attempt to meet special student needs in a cost-effective way.

Your supervisor at Mammoth Data Systems—Jane Gray, manager of new product development—asks you to prepare a draft of the proposal. The document is important to her because of the potential for involvement of her department in developing the new systems. The proposal will also be reviewed by her superior, Ariel Caliban, vice president of small systems and networks, before it goes to Dr. Prospero, president and founder of Mammoth.

Dr. Prospero's development of the Pro-9000 computer gave the company its start in the early 1960s. Though Mammoth's earnings and market share have declined in recent years, Dr. Prospero remains a champion of mainframe-oriented solutions. He and Mr. Caliban have had many "friendly" arguments about the virtues of large centralized systems versus PC-based local area networks.

1. For your company, what is the purpose of the proposal you are asked to draft?

2. What additional purposes are important to Jane Gray, Ariel Caliban, and President Prospero?

3. Name as many segments of the audience for your proposal as you can.

4. Given your audience(s), what should the general language level of the proposal be, and how should it handle technical detail?

5. What are the motivations, interests, and pressures felt by (a) school board members, (b) teachers, (c) parents? Is there a common ground for all these groups that can serve as a basis for your proposal?

THE DEVELOPMENT PROCESS

So how do you go from the issues of subject, purpose, and audience to a finished report? Do it the same way you get most of your work done: Set up a process.

The process approach will break the task into manageable parts. It will let you refine and polish these parts over time. It will ensure that first things get done first. And it will help you avoid the false starts and wasted effort that cost you more time in the long run. It is a way of working smarter rather than harder.

The process you adopt can be modified for particular assignments. Your instructor or your supervisor may structure the process for you by requiring certain preliminary "products" (such as an outline or a rough draft) by certain dates. If you are working with a partner or in a group, parts of the process may be assigned to

FIGURE I–1
Basic stages of the development process.

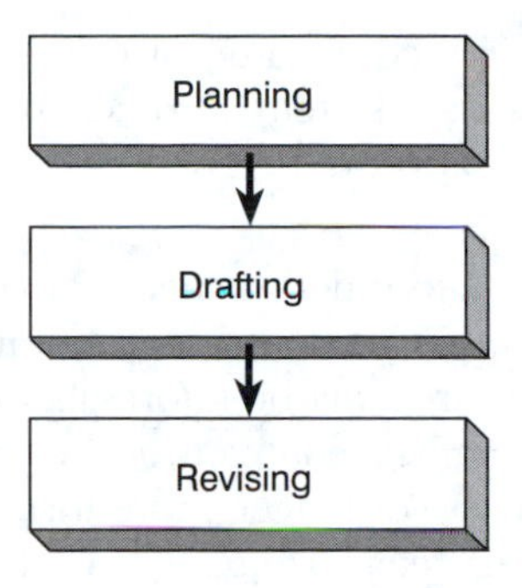

others. In its most basic form, however, the development process includes at least the three stages shown in Figures I–1 and I–2.

By allocating more time and effort to the first stage, you may reduce the time needed for the second. That savings can be profitably applied to the critically important third stage. As a rule of thumb, and subject to considerable variation, the time should be divided as shown in Figure I–2.

Let's take a closer look at each of these stages.

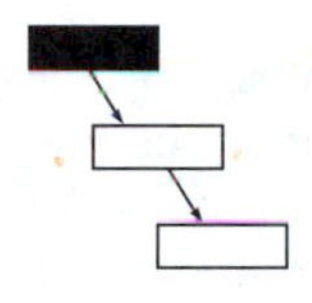

Planning

Key steps in the planning (prewriting) stage are as follows:

1. Clarifying the assignment
2. Research
3. Thinking
4. Formatting
5. Outlining

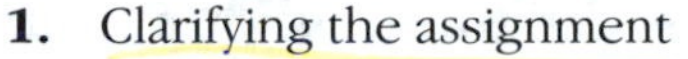

If you have been asked to write a report, start by *clarifying the assignment*. Review your notes, and check with the instructor or supervisor on parts of the task you're not sure of. If the idea for the report was your own, review your purpose and identify your audience. Then write a statement of these to guide your work. Here is an example for a technical sales department:

> Prepare a two-page report on wireless laptop computers for Andrew Chang. Recommend department purchase for all sales staff, emphasizing

Planning (35%)	Drafting (15%)	Revising (50%)

FIGURE I–2
Time allocation in the development process. Time spent planning and revising is most productive, and reduces the length of the drafting stage.

> the practical value of these machines for demonstrating capabilities of our data systems. Expect that Andrew will copy the VP of marketing to seek approval.

Next, consider the need for *research*. This is a step that is often divided among members of a group in a collaborative project. Whether shared or done individually, much research will be informal, spontaneous, and focused on a few key questions: people's names, company addresses, model numbers and specifications, costs, and similar data. For longer, more formal reports, you may need to obtain exact citations of sources, exact language, tables of statistics, and even entire manuals to be attached as appendices.

For the informal research, let the need dictate the source: Ask a colleague, visit a company or department, check your own files. For other projects—for example, to establish the state of the art in a technical field or to determine trends—you may need to do more systematic research. In that case, go to the library and use the available print resources or electronic references to guide your search. If you have access to the Internet, you may use some of the common search software (e.g., Gopher, Mosaic, Archie, WAIS) to identify and retrieve data relevant to your project. (See Unit 5.1, Research, for additional discussion of search strategies.)

Now, take a little time to *think* about your subject. This includes reviewing the assignment and the questions of purpose and audience. Try brainstorming (generating a free flow of ideas): Write your points down as they occur to you, in the order they come. Try focused brainstorming: all the benefits you can think of to a proposal, for example, or, for your progress report, all the projects completed last month.

Make lists, think on paper (or on the word processor), record everything, reject nothing. When you have listed all the points you can think of, go back and start clustering related ideas, throwing out repetitions and irrelevant points, and fleshing out undeveloped points. This process (see Figure I–3 for an example) will lead you toward the task of organizing your project.

Sketch out a *format* next. The format is a rough outline of your report, including the sequence of its major parts. While there is general agreement on the formats of common kinds of reports, your instructor or your company may also require that you use a specific approach adapted from the general format. When the format has not been specified, refer to the units in this book for descriptions and examples.

In the rental car case in Figure I–3, a brief recommendation report might be transmitted on a memo form. In addition to the formal orientation statements *(To:, From:, Subject:, Date:)* at the top of the memo, format would include your general organization of the material. Which of the clusters of factors in Figure I–3 would you consider most important, for example? In a memo, you would probably want to discuss these factors first and follow with less important material. (For a closer look at memos, see Unit 3.1.)

After you decide on the general form of your report and the general order of material, develop an *outline* of the project. Fill in the major sections of the format with the ideas, facts, and data you generated earlier.

FIGURE I–3
(a) Brainstorming of criteria for selecting a rental car agency for sales staff. Parenthetical items were added after first listing, and "durability" was eliminated later as a nonmaterial factor, since cars would be new; (b) a first version of the clusters of factors; (c) criteria grouped into clusters.

A. Rental Car Factors
- Cost
  (Daily vs. long-term rates)
- Mileage
- ~~Durability~~
  ~~(dependability)~~
- Size
  (head/leg room)
  (passenger room)
- Discounts for # of cars
- Insurance
  (collision)
  (vandalism)
- Four vs. 2 doors
- Safety features
  (airbags)
  (antilock brakes)
- Suited for long-distance driving
- Comfort
- Good in bad weather
  (traction)
- Off-road driving
- Accessories
  (Minor: cup holders, coat hangers)
- In-dash CD players
- Mileage/route trackers
  (i.e., trip computers)

B. 1. COSTS
2. PERFORMANCE
A. ROAD
B. BUSINESS
3. COMFORT
4. SAFETY

C. Rental Car Company Selection Criteria
(Clustered List)

1. Costs
   - Rates
     — Daily
     — Long-term
     — Volume discounts
   - Mileage
   - Insurance
     — Collision
     — Vandalism
2. Performance
   - Driving
     — Traction
       — Bad weather
       — Bad surfaces
     — Smooth ride (for distance)
   - Business productivity
     — CD player
     — Trip computer
3. Comfort
   - Size
     — Head/leg room
     — Passenger room
   - Four vs. two doors
   - Seat configuration
     (for distance driving)
   - Accessories
4. Safety
   - Airbags
   - Antilock brakes
   - Performance factors also figure here

Here's how it might work in another case. Suppose you want to write a job application letter in the following format:

Paragraph 1.
Introduction
- **a.** Interest-arousing opening
- **b.** Position applied for

Paragraph 2.
Qualifications for position
- **a.** Education
- **b.** Experience In a very general way, with perhaps one tantalizing detail

c. Skills/abilities
d. Personal qualities

Paragraph 3.
Close
a. Refer reader to resume
b. Facilitate response

While this format serves as a general guide to your use of material, it doesn't tell you what specifics to use or how to use them. What would you do to "facilitate the response" of the potential employer, for example? To help you plan for specifics, you might prepare a *working outline.* Use the points you listed in the brainstorming stage.

Thus, a working outline for the second paragraph of the application letter might look like this:

2. Qualifications (features/benefits)
   a. Education—B.S., Bradley U.; A.A.S., Tarrant JC
      i. Courses
      ii. Projects related to position
   b. Experience
      i. Faculty Asst., Bradley (lab support)
      ii. UPS Adm. Asst. (systems, office automation)
   c. Abilities, traits (examples = proofs)
      i. Design skills—Bradley project, 1st place
      ii. Leadership—UPS example
      iii. Work with people—FA, Bradley

Compared to a formal outline (see Unit 5.2, Outlining), the working outline is less structured but has more of the detail and strategy that will be used in writing the rough draft. In the section on abilities and traits, for example, the writer reminds himself to use the strategy of supporting claims with concrete examples, and prepares to use an example suggesting leadership traits demonstrated at a former job.

The working outline serves as a script for the next major stage of the process.

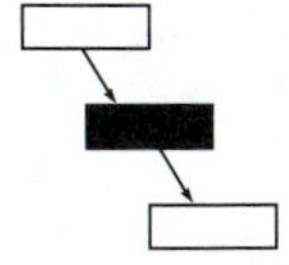

Drafting

Having laid the groundwork through your planning activities, you should now be ready to write a *rough draft* without spending excessive time and effort on it. If you remind yourself that you are producing a first, approximate version of the finished product, the writing will flow easier. Remember also that you will be doing serious revision and polishing in the next stage. You will do a much better job of clearing up problems and filling gaps at that point.

Here are some suggestions for writing a first (rough) draft:

1. Do it in one sitting, while your general plan is fresh in your mind.
2. Write from your working outline.

3. Write rapidly, and don't let yourself bog down in details or word choices.
4. If you can't think of a word, leave a blank space and go on.
5. Keep all your planning data and ideas in front of you for quick reference.
6. Double- or triple-space your lines so you can add material later.
7. Write on a word processor if you can. This will help you edit and revise later. You also won't have to struggle with spacing or shifting of material.

Notice how many of these suggestions concern *getting through* the draft stage so that you have a "complete" text to work with. Of course, the text is complete only in the sense of having reached an end, somewhere after a beginning and a middle. Yet, seeing an end to the project is a major advantage. Gaps and imperfections are easier to fix if you see the surrounding material and know where the entire structure is taking you.

Don't make the common mistake of spending most of your time wrestling with a first draft. Give it about 15% of your time, and go on.

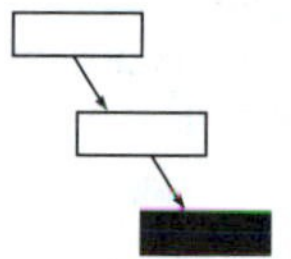

Revising

When we say that someone is a good writer, it would probably be more accurate to say good editor or good reviser. Much of the effectiveness of good writing is created in revision, not in the first effort.

Ernest Hemingway, a writer much admired for his clear, simple sentences, vivid descriptions, and powerful emotional effect, revealed something of how he worked. He said he wrote every morning till midday while he felt fresh. He produced a page or two on a good day, a few paragraphs more typically. Most of his time was spent rewriting sentences that didn't work to his satisfaction the first time, or the second, third, or fourth time. His simple, powerful prose, it seems, was the result of constant revision, not of sudden inspiration.

Some of the strategies you can bring into the revision stage include the following:

- ☐ Peer review
- ☐ Correcting
- ☐ Ensuring completeness
- ☐ Restructuring
- ☐ Improving style
- ☐ Adding graphics or visuals

Working against a deadline, you may not be able to do all of these, or give due attention to each. They are examples, nonetheless, of the many available ways to improve your first efforts.

Peer Review. You may have heard this advice before: "Put a first draft aside for a few days; then come back to it with a fresh eye." Good advice, but not too helpful if your paper is due the next morning.

A more practical way to gain that fresh viewpoint is to ask a friend or co-worker to read your draft. If you can, choose someone with a knowledge of your subject. Make a copy of the paper, and ask your reader to jot comments right on the copy.

In collaborative projects, peer review is integral to the process as group members adjust each other's efforts toward a common outcome. With advances in electronic communication, groups no longer need to be physically together to conduct peer review. Internet users with software that provides a *whiteboard service* can view a common document from remote sites and change or add text or graphics to the draft. The Internet also offers assistance in the form of Online Writing Labs (OWLs) established by colleges and universities to answer writers' questions or make suggestions for change.

Reviewers can give you many kinds of feedback—about the clarity and correctness of your writing and about the quality of your ideas. You can also focus the review by asking a reader to concentrate on certain features and parts of the draft.

If the language of the draft is still rough, ask the reader to concentrate on the structure and content instead. If you've done some polishing already, ask for comments on readability as well.

Treat the feedback you get seriously. Even if you don't agree that one of your paragraphs is "confusing," for example, remember that the reader is serving as a trial audience for you. In telling you whether your message is getting through, "the audience is always right." Go back and fix the problem.

Corrections/Completeness. Check both the data you have used and the way you incorporated data into your report. At one level, you might look simply for errors in transcribing figures and information. If you've done calculations, double-check these for accuracy. If you obtained data from a single source, look for confirmation in other sources as well. In addition to using peer reviews, search further yourself. Many technical and scientific articles, for example, provide a listing of *their* sources, which may include other articles with their own citations. Checking sources puts you in touch with a widening circle of researchers and experts to help strengthen your conclusions.

Check also that your development of material is logical, consistent, and complete. If you have left out an important assumption or qualification, for example, your conclusion may sound unconvincing or illogical to the audience. This is your chance to strengthen your case with additional evidence, information, and ideas.

Restructuring. You can reorganize your material on several levels and in several different ways.

At the simplest level, you might find that your presentation is improved if you move material from one place to another. For example, a list of benefits to a proposal might be moved from the end to the beginning of a section, to serve as an overview or preview rather than a summary.

You can also reorganize in a more general sense by moving major sections so that your structure follows an order-of-importance or a top-down strategy. Or you might reorder material into a chronological or reverse-chronological sequence, or according to a functional or spatial arrangement. (See Figure I–4 for a discussion of organizing principles.)

ORGANIZING PRINCIPLES

- *Order-of-importance*
- *Reverse chronological*
- *Order-of-climax*
- *Functional*
- *Chronological*
- *Spatial/Physical*

Each of these organizing principles can serve your needs best in certain situations. The order-of-importance strategy helps gain more attention at the beginning and may work well with skeptical or impatient audiences. Its counterpart, the order-of-climax, *builds up* to the most important point and may achieve a stronger cumulative effect, provided you can keep the audience interested.

A chronological (order-of-time) approach is effective for showing development, trends, or a chain of causes and effects over time. A reversed chronology is used when recent events matter more than earlier ones—as when a list of prior jobs is assembled on a resume or prior residences are listed on a bank loan application.

A functional organizational approach lets you divide the subject by tasks done or effects achieved. This approach works well in describing complex organizations, systems, or mechanisms by subdividing the subject according to the functions performed by the members or parts of the whole.

A spatial/physical organizational approach might help achieve visualization through detailed physical description, indicate shape and size, point to location, and obtain similar results that may be important as *part* of many technical and business writing situations.

FIGURE I–4
Ways of organizing may vary with the subject, purpose, audience, and situation.

If you are working under a specified format, check that your material fits these requirements. If the format is of your own choosing within only general guidelines, consider whether your arrangement best serves your purposes.

Both major and minor restructuring will be greatly facilitated by your use of word processing technology. Your ability to restructure material is one of the strongest recommendations for adopting this approach. (See Unit 5.5 for document design strategies using word processing techniques.) Even with traditional paper-and-pencil methods, however, restructuring is still feasible and well worth the effort.

Improving Style. The major considerations of technical style are discussed and illustrated in the next section, pages 15–28. Spend some time studying this section and then refer to it as the need arises. In revising your rough draft, focus on a limited number of elements that can make a marked difference in effect.

Sentence length and variety might be one of these elements. Look for opportunities to divide longer sentences into several shorter ones; at the same time, make the transitions between sentences and paragraphs more emphatic to help the flow of ideas.

A basic element of effective style is clarity. Look for passages that your peer reviews or your own rereading indicate need to be clearer or less ambiguous. One way to improve clarity is to tighten the reference of all personal and indefinite pronouns and to make sure that modifying phrases or clauses point to the intended words. (See Figure I–5 for help in eliminating reference problems.)

Another way to improve clarity is through greater precision, something you can achieve through additional detail or more specific language. The specific term is preferred because it will help you improve conciseness as well as convey meaning more accurately. Among other goals of effective technical style, look for the shortest, simplest, and most direct constructions to convey your meaning.

How much polishing you do depends on the need and on your willingness to hold out for a better way to say something. The difference you can make is substantial.

Adding Graphics or Visuals. Add graphics or visuals as a last step, after the surrounding material is largely finalized. It is likely that opportunities for use of visuals would have occurred to you during earlier stages. At this later point you can check

CHECKING REFERENCES

Unclear reference of pronouns or an unclear relationship between modifiers and their antecedents may add to confusion or ambiguity in sentences. You can revise a first draft to make sure that pronouns and modifiers point directly to the intended words:

Unclear Reference	***Improved Reference***
1. The lawyer told his client that *he* would be responsible for research costs.	1. The lawyer demanded that his client accept responsibility for research costs.
2. The judge instructed the jury to disregard the defendant's service in the Reagan administration during the early eighties. *This* upset opposing counsel.	2. Opposing counsel were upset that the judge had called the jury's attention to the defendant's high rank in the Reagan administration. *Or:* . . . that the judge had chosen to suppress the defendant's service in the controversial Reagan administration.
3. The witness contradicted earlier testimony, *casting* doubt on the defendant's alibi.	3. The witness contradicted earlier testimony, which had cast doubt on the defendant's alibi. *Or:* . . . Contradicting earlier testimony, the witness cast doubt on the defendant's alibi.
4. *Summoning* both counsel to her chambers, a brief recess was called by the judge.	4. Summoning both counsel to her chambers, the judge called a brief recess.

FIGURE I–5
The reference of pronouns or modifiers may need to be clarified during the revision stage.

whether these ideas still apply and look for new opportunities resulting from reorganization, rewording, or simply reconsideration on your part.

A good place for illustrations is a passage that includes detailed data, process description, relationships, patterns, and other material that is hard to describe in words alone. In addition to clarity, you may seek to enhance dramatic effect or persuasiveness or variety through visuals and graphics.

Selecting graphics and then blending them into your material involve many considerations that must be weighed against your strategies and purposes. See Units 1.8 and 4.6 for additional information.

TECHNICAL STYLE

Style in technical writing is based on several elements common to all writing. These key elements include sentence length and complexity, word choice, and tone. In revising their drafts, technical writers should work to shape these elements for maximum readability.

General Characteristics of Style

Sentence Length. Average sentence length affects the reading level of a passage. Usually, shorter sentences (10 to 15 words) make the writing easier to read and understand. Although effective writers use various sentence lengths to avoid a monotonous effect, they also hold down *average* sentence length. Longer sentences (20 words and up) slow down communication because they also tend to be more complex.

Sentence Complexity. Sentence complexity is largely a function of sentence structure. Structurally, a sentence becomes more complex when it has several clauses that depend on each other and when the completion of the meaning is delayed or interrupted by other clauses or phrases. Three basic sentence types can be used:

1. *Simple sentence:* one independent clause

   Divers entered the hull to assess the damage.

2. *Compound sentence:* two (or more) independent clauses joined by a conjunction

   Divers entered the hull, and they assessed the damage.

3. *Complex sentence:* one independent clause and one or more dependent clauses

   After the tanker struck a reef, divers entered the hull to assess the damage.

Most readers can handle sentences like the preceding with little strain. These sentences are relatively short, and even the complex type is not difficult. However, if the complex version had additional details and explanation, the result might be a *complicated* sentence:

Poor: After the tanker struck a reef, one that had not been charted accurately, and which went undetected because of a malfunction in the sensing equipment at the critical time, divers entered the hull to assess the damage.

The preceding sentence not only runs to 37 words but delays the completion of the thought with additional detail. Sentences like these may be divided into several parts to improve readability:

Better: After the tanker struck a reef, divers entered the hull to assess the damage. The reef had not been charted accurately and went undetected because of a malfunction in the sensing equipment.

Readability and interest can also be increased by varying the structure of sentences in a passage. You can choose among the three basic sentence types, and you can vary the word order within a sentence. Transitional words can also be used to link sentences while creating variety. The complex sentence usually offers a choice: You can put the independent clause before the dependent clause or after it:

I shut the motor off as soon as the vibrations began.

As soon as the vibrations began, I shut the motor off.

Notice how the following paragraph blends short and long sentences, a variety of sentence structures, and effective transitional devices:

> Lorenz asked a second question. Suppose you could actually write down the complete set of equations that govern the weather. In other words, suppose you had God's own code. Could you then use the equations to calculate average statistics for temperature or rainfall? If the equations were linear, the answer would be an easy yes. But they are nonlinear. Since God has not made the actual equations available, Lorenz instead examined the quadratic difference equation.[2]

Word Choice. Word choice is a third important contributor to readability. As in the use of sentences, shorter and simpler words make a passage easier to understand. Elaborate, multisyllable words do the opposite.

We wanted to limit emissions *originating* from the operation of plants *utilizing* coal-fired boilers in order to *minimize* damage to the environment.

Could we replace the underlined words with shorter, simpler ones without changing the meaning?

2. James Gleick, *Chaos: Making a New Science* (New York: Viking Penguin, 1987), 168.

1. *Originating:* Replace with *arising* and save two syllables. Or better yet, eliminate the word.
2. *Utilizing:* Replace with *using;* save two syllables.
3. *Minimize:* Use *reduce?* No, because minimize means "to reduce to the smallest amount possible."

Words should be chosen for their meaning. When meanings do not differ, choose a short, simple word over a long, elaborate one. Use the longer word if it is more exact, but don't sprinkle your writing with long words just for effect.

Tone. The tone of the writing expresses the writer's attitude. A business letter to a complaining customer can sound hostile or conciliatory. A technical report or memo can convey a friendly or impersonal tone; it can be formal or casual; it can use humor and irony; it can even express modesty, pride, or enthusiasm.

All the elements of style previously discussed contribute to tone, particularly word choice. Writing that uses elaborate words for simple ideas creates a stuffy tone rather than a dignified or learned one. Compare the following two versions:

Stuffy: It is imperative that a verification of each student's status with respect to his or her presence in the classroom be conducted at the inception of each academic session.

Plain: Attendance must be taken at the start of each class.

The snooty tone of the first version comes not just from word choice but also from its excessive length and complexity.

Other varieties of tone also might be expressed in this statement:

Apologetic: I know it's a lot of trouble, but could you please take attendance for each class?

Satiric: To keep the bureaucrats off our backs, you'd best take attendance for each class.

Informal, casual: Just take a head count in each class.

Any of these may be suitable in certain situations—between friends, for example. The stuffy version would be hard to justify in any circumstances.

As mentioned, the general elements of style—sentence length and complexity, word choice, and tone—are factors in technical writing as well. Technical writing, however, also places a greater emphasis on some of these elements and adds unique features of its own.

Characteristics of Technical Style

In technical writing, the following qualities are important or even essential:

- ☐ Objectivity
- ☐ Precision
- ☐ Clarity
- ☐ Economy
- ☐ Audience awareness

Although much good writing, including fiction, is marked by precision, clarity, and economy of means, objectivity is central to technical and scientific writing. Furthermore, because of the specialized subject matter of technical writing, audience awareness places special demands on writers to choose the appropriate level of language.

Objectivity. *Objectivity* means neutrality—the absence of bias. Objectivity suggests an attitude and approach to one's material that is expressed in being factual. It is an attempt to present things as they actually are.

In contrast, subjectivity introduces personal perception and often adds emotion as well. The subjective viewpoint presents things as they look when filtered through personal experience, values, and attitudes.

The use of objective and subjective approaches results in two kinds of meaning: denotative and connotative. The denotation of a word is its neutral meaning, as given in a dictionary. The connotation is its subjective, personal meaning, as given by someone who feels strongly about the subject one way or the other. Many things, therefore, can be described by a range of terms within three main categories (see Table I–1 and Figure I–6).

The objectivity of technical style is expressed through terms like those in the middle column in Table I–1. This does not mean that effective technical writers never engage in judgments or express emotion, just that they don't prejudge their subject. A technical report on nuclear weapons may very well conclude that they are a threat, but this conclusion would likely be preceded by an objective evaluation of destructive capabilities.

TABLE I–1
The Spectrum of Meanings

Positive Connotation	Denotative Meaning (Neutral Term)	Negative Connotation
public service organization	public organization	bureaucracy
financier	investor	speculator
information management	data processing	number crunching
educator	teacher	pedant
thrifty	economical	cheap
financial plan	budget	fiscal straitjacket
executive	manager	boss
prudent	cautious	timid
bold	adventurous	reckless
wonder drug	patent medicine	quack remedy
marketing	selling	peddling

FIGURE I–6
Connotation may be positive or negative.

Good technical style also does not require "impersonality" of the kind that avoids all references to the first person *(I, me, we, us)*. A lab report may remain objective even though it says, for example:

I measured the Q_1 base-to-emitter voltage.

rather than

The Q_1 base-to-emitter voltage was measured by the experimenter.

If the first-person version is accurate, as well as shorter and more direct, it is preferred to the passive, indirect, third-person construction.

Nor is the writer's objectivity marred by humor and other devices of lively writing. Objectivity does not require a grim, mechanical treatment of the subject. A lively tone can also be achieved in writing that has a serious purpose and an impartial spirit.

Precision. Precision is a second important feature of good technical writing. Like objectivity, precision derives from effective use of facts and details to build one's case. Technical writing is precise when it has the following characteristics:

- ☐ It uses concrete language.
- ☐ It uses exact dimensions and units of measure rather than general estimates of size, weight, volume, frequency, intensity, and so on.
- ☐ It uses technical terms that have well-defined meanings within a field.

Concrete language is specific. It refers to particular members of categories rather than the broad categories themselves. It is the opposite of *abstract* language, which relies on general terms that lack detail. If I refer to a high-backed dining room chair with padded armrests, I am being far more concrete than if I speak merely of furniture, home furnishings, or possessions. I am staying low on the ladder of abstraction (see Figure I–7), where things have specific size, shape, and use.

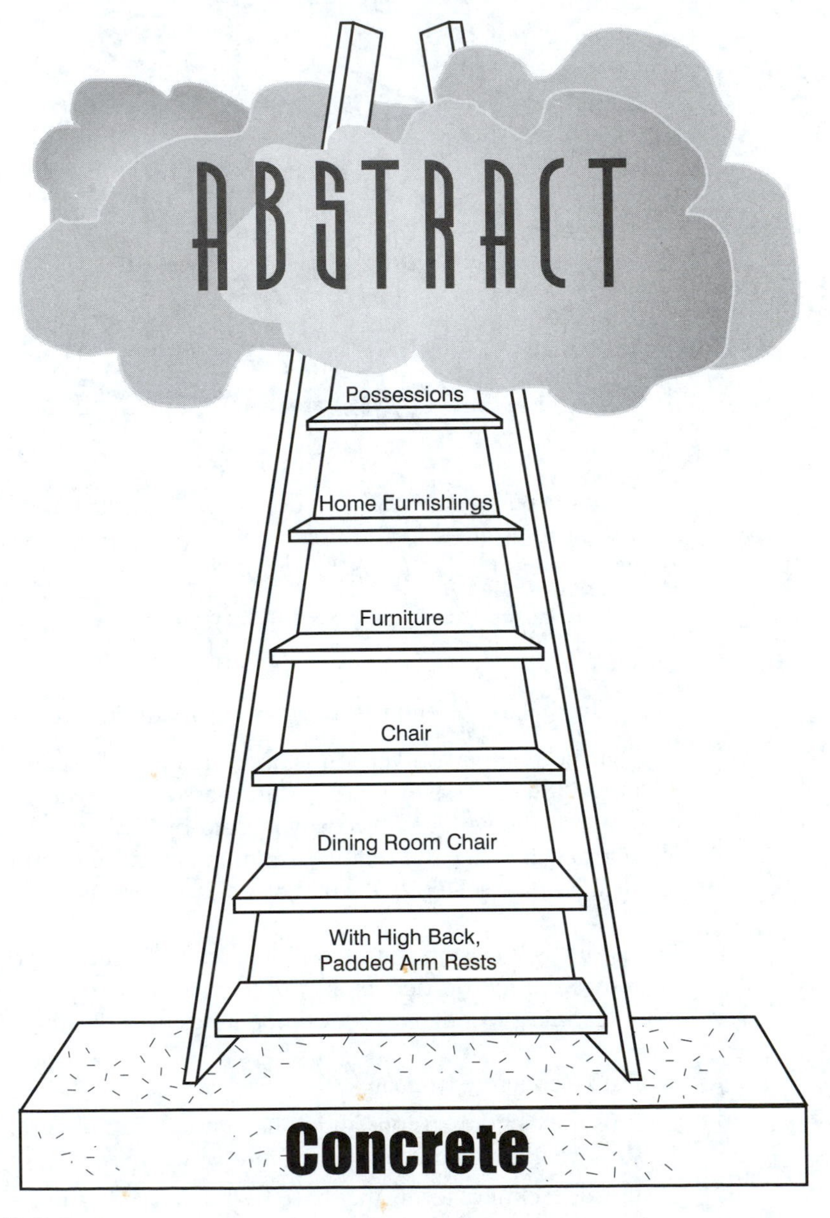

FIGURE I–7
The ladder of abstraction.

I could go lower on the ladder yet and refer to the chair's height, color, texture of materials, and many other specific details. How specific I get depends on my writing purpose. The level I selected is specific enough to tell a repairman which chair to fix, for example, but it would not provide enough detail to instruct a furniture maker on what kind of chair to build for me. Technical writing also deals with abstractions (*efficiency*) and abstract terms (*equipment*), but more often it relies on specific kinds of equipment, characteristics, and processes.

In the same way, exact *units of measurement* are important in technical reports. Readers not only need to form a clear picture of the object or process being described but may also need to install, service, repair, analyze, or evaluate it. For such purposes it is not enough to call the Whiztech 886 a "high-speed" protocol analyzer, for example; readers need to know that it can analyze data streams at up to "72 million bits per second."

Along with exact units, *technical terms* help build precision. Such terms have definite meanings within their field and when used properly can sharpen a description or discussion. Proper use, however, includes awareness of the reader's knowledge. Because some readers may lack the technical background of the writer, it is good practice to define technical terms as they are introduced.

Clarity. Clarity is important to all writing because it underlies most other purposes. In user's manuals, instructions, or specifications, clarity is critically important to ensure proper use and avoid possible damage or injury. Word choice is a major contributor to clear writing, as are the techniques for achieving precision. Two other objectives should be pursued for the sake of clarity: (1) completeness and (2) correctness.

Complete information and explanation require a mind-set similar to the journalist's, whose newspaper article answers the key questions of the reader: *What? Where? When? Who? Why?* and *How?* Not all these questions are pertinent in all cases, but the total set provides a framework for complete coverage.

Correct usage, punctuation, and grammar also help ensure clarity. When these elements are handled correctly, they become "invisible" by effectively supporting the writer's meaning without getting in the way. Errors in correctness, however, can steal attention and obscure the meaning.

Misplaced modifiers—words or phrases that point to the wrong part of the sentence—are particularly distracting to the reader:

Distracting: Training sessions will be offered at several sites *covering the operational characteristics of the equipment.*

Clear: Training sessions *covering the operational characteristics of the equipment* will be offered at several sites.

Distracting: *By using programmed sequences of instruction under computer control,* individualized rates of progress can be achieved by students.

Clear: *By using programmed sequences of instruction under computer control,* students can achieve individualized rates of progress.

Careless use of punctuation can also distract and confuse:

Confusing: The proposal which deals with administrative computing was discussed at today's meeting.

(Had several proposals been submitted or just one?)

Clear: The proposal, which deals with administrative computing, was discussed at today's meeting.

Clear: The proposal dealing with administrative computing was discussed at today's meeting.

Confusing: Several members were absent from the meeting fortunately, the key agenda items were rescheduled.

Clear: Several members were absent from the meeting; fortunately, the key agenda items were rescheduled.

Economy. Economy of language is achieved through efficient use of words, phrases, and sentences. It means the fewest words for the desired meaning. In technical writing, shorter is decidedly better because it helps the reader understand and retain complex material.

Concise expression is not easy to achieve in the first effort, when attention is focused on what to say rather than how to say it. But first drafts can be dramatically improved by careful editing. Following are some ways to make writing more concise.

1. Delete unneeded words or phrases.
2. Substitute single words for phrases.
3. Avoid *there is, it is* constructions.
4. Limit use of passive voice.
5. Revise indirect sentences into direct forms.

Thus, *unneeded words* are redundant; they restate a meaning that has already been expressed, as in *repeat again* or *surrounded on all sides.* In Figure I–8 the italicized words can be cut out with no loss of meaning.

Wordy phrases can be replaced by single words, which are particularly effective substitutes when they are action verbs (see Figures I–9 and I–10).

Avoid constructions that begin with *there is, there are, it is, it was,* and similar forms. These delay introduction of the subject and add words that don't advance the meaning.

Wordy: There are three assumptions on which this proposal is based.

Concise: This proposal is based on three assumptions.

Wordy: It is by the use of subroutines that we can substantially reduce programming time.

Concise: We can reduce programming time substantially if we use subroutines.

FIGURE I–8
Redundant expressions.

actual experience
advance planning
assemble *together*
basic fundamentals
blue *in color*
close proximity
collaborate *together*
complete monopoly
completely eliminated
consensus *of opinion*
elliptic *in shape*
end result
expensive luxury
few *in number*
first began
filled *to capacity*
final outcome
mutual cooperation
necessary prerequisite
original source
past history
future plans
personal opinion
the reason is *because*
revert *back*
small *in size*
throughout *the whole*
true love

Wordy: There was no need for an additional data link.
Concise: An additional data link was not needed.

Passive voice constructions are longer and, as their name implies, weaker than active forms. In the active voice, the subject performs the action; in the passive mode, the subject only receives the action. Notice the contrast in length and energy level between the following pairs of statements:

FIGURE I–9
Wordy phrases versus action verbs.

Phrases	*Verbs*
bear in mind	remember
bring about	cause
afford an opportunity	allow
are of the opinion that	think/believe
come into contact with	meet/touch
give consideration to	consider
has the ability to	can
has a need for	needs
held a meeting	met
inquired as to	asked
made a decision	decided
make application to	apply
produced an improvement	improved
reached the conclusion	concluded
takes exception to	disagrees

Passive: Many of these tasks *are* now *performed* by office computers.
Active: Office computers now *perform* many of these tasks.
Passive: Each operator's output *is monitored* by the supervisor.
Active: The supervisor *monitors* each operator's output.

The passive voice often hides or obscures the performer of the action:

Passive: Your suggestion *is being evaluated.*
Active: Finchley *is evaluating* your suggestion.

Passive voice is appropriate, however, when the performer of the action is unknown, or when the performer is less important than the action itself.

FIGURE I–10
Other wordy phrases.

Wordy	*Concise*
a large number of	many
as a general rule	usually
as far as . . . is concerned	about
at the present time	now
at that point in time	then
at such time as	when
by means of	by
due to the fact that	because
for the purpose of	for
in addition	also
in all probability	probably
in length/height	long/high
in many/most cases	often/usually
in order to	to
in regard to	about
in spite of the fact that	despite
in the event that	if
in the near future	soon
in the vicinity of	near
of the order of magnitude of	about
on the basis of	by, from
prior to	before
subsequent to	after
through the use of	by, with
until such time as	until
with reference to	about
with the exception of	except

Active: Somebody vandalized the Oakridge Service Center last night.

Passive: The Oakridge Service Center was vandalized last night.

Finally, just as the fastest way to reach a destination may be a straight line, *direct sentence structure* is more efficient and easier to follow than indirect forms. Direct sentences follow many of the guidelines for economical language. The following techniques are most important to writing direct sentences:

- ☐ Use the active voice.
- ☐ Use single words for wordy phrases.
- ☐ Avoid redundancies.
- ☐ Avoid *there are, it is* constructions.
- ☐ Minimize interruptions and delays in completing the meaning.

Indirect: In an article that was published in *Computerworld,* there is an explanation of this phenomenon that has been so puzzling to so many.

Direct: A *Computerworld* article explains this puzzling phenomenon.

Indirect: Under the verification procedures, it is required by the commission that manufacturers of equipment that is subject to verification must submit a report detailing the measurements that have been made by them to determine whether they are in compliance.

Direct: The commission's verification procedures require manufacturers of certain equipment to report the measurements they have made to determine compliance.

Indirect: In spite of the voluminous statistics that were included in the proposal, due to the fact that many of these figures were drawn from sources of questionable reliability, the proposal's recommendations were rejected by the executive committee.

Direct: The executive committee rejected the proposal because its voluminous statistics seemed unreliable.

Adaptation to the Audience. Adapting material to the audience is a final important requirement of good technical style. In its simplest sense, this adjustment responds to the technical level of the readers. The level might be set by one of the following:

1. A *general audience* with no significant command of the technical principles or terminology of the subject field
2. A *technician-level* audience familiar with the operational details of the technology but not with its theoretical basis
3. *Expert* readers, such as engineers and scientists, who may know the theory better than the practical details

Two other types of audiences should be considered. For these readers, the technical level may not be the prime consideration.

4. The *executive* readership, concerned mostly with the business implications of the report
5. The *mixed* audience, which may include various technical levels and various business, social, and political concerns

How should a writer adapt to these possible audiences? For *all of them,* the most effective style is the clear, precise, objective, economical writing described earlier in this section. In addition, you should pay special attention to the *vocabulary* and the *organization* of the report. Following are some strategies for adapting your writing to the five audiences listed.

1. *General audience* (for example, end users of technical devices or products): Identify or define technical terms when first used. Make sure definitions are not circular. Use graphics to support explanations. Focus on what's important to the reader, that is, how to use the product.

2. *Technicians:* Technical language about operations and service details will be familiar to these readers, but theory may need to be explained. Focus on the reader's interests: the construction, installation, servicing, and testing of the equipment. Use graphics to support explanations.

3. *Experts* may be primarily interested in the design or evaluation of the technology. They may also be more comfortable with it in schematic or mathematical form and may need more guidance at the hands-on level.

4. *Executives* will likely be more interested in the impact on costs, production requirements, personnel issues, and company politics than in the technical aspects of the report. Because of time constraints, they may devote most attention to the summary and the conclusions, and skim the technical discussion. For these readers, focus on economic and other benefits but don't neglect the technical analysis, which the executive may pass on to an "expert" reader for an opinion.

5. *Mixed* audiences of general readers, executives, and several kinds of technical readers may be common for reports and proposals within an organization and even more likely for documents that go outside. Rather than writing such reports to the lowest common denominator, begin with a summary in clear, nontechnical language, and then develop the rest of the report in more depth and technical detail. Alternatively, you can keep the body of the report relatively simple, and reserve the technical discussion for an appendix.

Most important, think of the reader's entire situation, not just the issue of technical expertise. How will the report be used—for information only or as a basis for action? What are the political pressures and vested interests that affect the reader? What do you know of the reader's attitudes and opinions about your subject?

Let's say that you intend to recommend the purchase of some expensive equipment that will save money in the long run. You also know that Vice President Finchley is facing severe pressure to reduce current expenditures. In this case, you may decide not to begin your proposal with the recommendation; it might be better to build your case gradually by showing how inefficient (i.e., costly), the present equipment is and then present your recommendation.

A Checklist for Effective Technical Style

Remember that an effective style in technical and other kinds of writing is one that makes the reader's job easier. To enable the reader to grasp your material easily, you have to work harder yourself. The most productive work of this sort you can do is to *revise* your first efforts. Leave yourself time for revision, and then be as ruthless as possible in pruning, reorganizing, and otherwise clarifying your writing. Ask yourself questions like those in Figure I–11.

FIGURE I–11
A checklist for effective technical style.

Yes	*No*	
☐	☐	Are my sentences generally short and uncomplicated?
☐	☐	Are my sentences varied in structure and length?
☐	☐	Are my words as short and simple as they can be, given my message?
☐	☐	Is my tone suited to my purpose?
☐	☐	Am I presenting things factually and impartially?
☐	☐	Am I using specific terms and precise dimensions when these are needed?
☐	☐	Am I giving complete information?
☐	☐	Am I using the language correctly, to clarify instead of confuse my meaning?
☐	☐	Can I cut out words and reduce phrases to single words?
☐	☐	Am I using constructions that are active, direct, and efficient?
☐	☐	Are my technical terms appropriate to my readers?
☐	☐	Have I shaped my presentation to suit the reader's priorities and pressures?

Technical Style Exercise

1. Revise the following sentences as appropriate to eliminate long, elaborate words and inefficient phrases in favor of shorter, simpler terms.
   - **a.** Our recently initiated system of facsimile terminals is considerably more rapid and less expensive, and it provides greater reliability than the utilization of either express mail or courier services.
   - **b.** Implementation of restrictive personnel policies on the part of the corporation was found to have exercised a significantly detrimental effect upon development of the characteristics of initiative and creativity as far as employees were concerned.
   - **c.** When I first came into contact with Dr. Chesterton, he indicated that he was of the opinion that an improvement could be produced in the experimental design of our survey.
   - **d.** At that point in time, due to the fact that the price of gasoline had risen considerably, we gave serious consideration to the purchase of a fleet of electric cars for the purpose of the delivery of orders to residential customers in our area of service.
   - **e.** In all probability, in spite of the fact that electronic monitoring systems are in place, until such time as dramatically increased security precautions are effected, a large number of the cash machines in the vicinity of the downtown commercial district will continue to experience tampering problems.
2. Replace the vague terms in the following sentences with concrete detail that fleshes out the meaning but does not change it.
   - **a.** Differences between the two sides were not entirely eliminated.
   - **b.** For a lot of reasons, it was a better buy for us.
   - **c.** Supplies are low; we need to do something about that soon.
   - **d.** The two events were related.
   - **e.** In time, the fortunes of the company improved.
3. Revise the following paragraph for better sentence variety. (Try to vary sentence length and structure.)

   Permanent magnets are made of hard magnetic materials. Such materials are magnetized by induction during manufacture. Residual induction makes them permanent magnets when the magnetizing field is removed. Permanent magnets may last indefinitely. They should not be subjected to high temperature. High temperature may lead to loss of magnetic properties. A temperature of 800°C makes iron lose its magnetic properties.
4. Review the lab reports you submitted in a technical course this term or in a prior term. Suppose you wanted to revise this report for submission in a technical writing or business writing course taught by a nontechnical instructor.
   - **a.** How would you change the language of the report?
   - **b.** What organizational changes, if any, would you make?
   - **c.** What elements, if any, would you emphasize and deemphasize compared to the original report?

A DEVELOPMENTAL CHECKLIST

Through practice and application, effective strategies for planning, drafting, and revising can become a natural part of the communication process, saving time and enhancing the results. Until these strategies begin to feel natural, a more deliberate, step-by-step approach may be advisable. A checklist such as the following can ensure attention to the important steps. (Copy the blank form first, for repeated use.)

A Checklist for the Writing Process

☐ The subject of my communication is ______________________________ .

☐ The general purpose of my communication is ☐ persuasive, ☐ informative, ☐ directive, ☐ analytical, ☐ evaluative, *or* ☐ a combination of several of these, namely ______________________________ .

☐ A specific statement of my subject and purpose in this communication is

____

____

______________________________ .

☐ My primary audience, or reader, is ______________________________ ; other members of the audience are ______________________________ .

☐ The level of technical understanding of the audience is as follows:

Primary audience: ______________________________ ;

Others: ______________________________ .

☐ The main concerns and interests of the audience are as follows.

Primary audience: ______________________________ ;

others: ______________________________ .

☐ The common ground for all members of the audience is the following as regards language and technical understanding: ______________________________

______________________________ ;

as regards motivations and concerns: ______________________________

______________________________ ;

☐ The following topics or issues need to be researched: ______________________________

______________________________ ;

possible sources for information on these topics are ______________________

____ .

☐ I have completed the necessary research.

☐ I have thought about the points I want to make, listing all ideas in an unrestricted order first.

☐ I have reviewed the initial list of ideas and ☐ clustered the related topics, ☐ amplified some topics, ☐ and discarded others.

☐ Thinking of my audience and looking at the revised list of topics, I have chosen the following organizational approaches.

a. The specific format type (if any) is ______________________ ;

b. the general organizing principle that best suits this material is

____ ;

☐ I have grouped the material into the following sections.

1. The beginning includes: ______________________

____

____ .

2. The middle includes: ______________________

____

____ .

3. The ending includes: ______________________

____

____ .

☐ Using the general ordering of this material, I have developed an informal outline by filling in the details of points I want to make.

☐ Working from my informal outline, and using its suggested strategies and examples, I have completed a first draft of my document.

☐ I have obtained one or more peer reviews of my first draft; reviewers were asked to focus on ______________________

____ .

☐ I have incorporated review comments into my revisions.

☐ I have checked claims and important data for accuracy and completeness.

☐ In reviewing the general organization of the document, I have moved or reordered material as follows: ____

____

____

____.

☐ I have improved the written style of the document by ☐ reducing sentence length, ☐ increasing sentence variety, ☐ improving word choice, ☐ strengthening transitions, ☐ clarifying pronoun reference, ☐ adding more specific detail, and ☐ using the following other methods: ______________________

____

____

____

☐ I have added graphics or visuals to help explain processes, relationships, patterns, and data.

HOW TO USE THIS BOOK

As summarized in the preceding checklist, this introduction is focused on a general strategy of development and revision. Most of the book, however, is organized around *units* describing particular kinds of writing or speaking projects.

The units that describe specific types of projects are self-contained and practical in orientation. The units contain a brief introduction pointing to the major purposes of that form of communication. They give examples of typical assignments or projects, both academic and career related. And they suggest an approximate length or range of lengths for the project.

Much of each unit is devoted to (1) an explanation and (2) an example of the writing or speaking project. The explanation frequently gives a suggested general format or organization scheme. Just as often, it reminds you that specific formats can differ from the suggested one.

Where appropriate, units also point to situations or scenarios that can affect the way communication is planned and carried out. In these cases, you are asked to think critically and choose the most effective approaches to your subject and audience.

The project explanations also offer guidelines for how to prepare these assignments. These suggestions provide specific applications of the process approach described in this introduction. You can use that approach by going through the checklist of planning, drafting, and revision steps for each assignment you receive. While shorter and better defined assignments may not require the full checklist, regular use will make the process second nature and help you handle longer, more ambitious communication projects.

Organization and Access

The table of contents shows that units are grouped into sections according to major communication types, strategies, and specific techniques.

The first section includes units on development methods that may be *part* of reports or presentations. The methods are not themselves reports, in other words, although they may be assigned separately for their value in helping you learn.

Major kinds of technical and business *reports* are grouped separately from *correspondence.* The latter, as evidenced by the growth in electronic transmission and retrieval, is an increasingly important form of communication that can benefit from specialized techniques and strategies.

Section 4, on *oral communication,* includes units on active listening, meetings, special applications of principles, and general guidelines for oral reports.

Section 5, on *research and development,* assembles a series of units on specialized techniques and issues that relate to the development process discussed previously. The final section of the book offers units on the general rules, forms, and procedures of writing. These last two sections serve primarily as reference tools.

Exercises are provided in the Introduction, in Technical Style; and at the end of Unit 1.8, Figures and Tables; Unit 3.1, Memos; Unit 3.4, Resumes and Application Letters; Unit 4.1, Active Listening; Unit 4.4, Meetings; Unit 4.5, Oral Reports; Unit 5.3, Using Sources; Unit 6.1, Basic Grammar; Unit 6.2, Vocabulary and Usage; Unit 6.3, Capitalization; and Unit 6.4 Punctuation. Answers to the first and the last four are at the back of the book. The checklist at the end of the Technical Style section may help you cover the important bases during the final stage of the development process. For a general review of developmental strategy, use the checklist on pages 29–31. Finally, an evaluation form at the end of Unit 4.5, Oral Reports, can guide your preparation for speaking assignments.

Your instructor may ask you to study parts of the book in conjunction with writing or speaking assignments. Use it for self-reference as well when you want to prepare a message or find practical guidance for a communications need. Use the table of contents to locate general categories of material, and check the index at the back for specific topics.

A Reminder

This book is a tool to help you develop your skills in a craft that must be practiced to be learned. Writing is the ultimate learn-by-doing craft. You can't learn to write by just reading about it.

Use this book to back up your planning, drafting, and revision. You can bring it in at any stage, but you will need to accept the basic challenge first: learn to communicate more effectively—learn by doing.

SECTION 1
Development Methods

Many of the reports described later in this book deal with subjects and purposes that require systematic presentation. Because the subject is technically complex or vital to an organization or includes safety considerations and liability issues, precise handling is required. The units in this section present structured methods for such basic presentation needs. The units include common methods, such as technical description or use of graphics, and common strategies that might be applied to *part* of a report, such as a set of instructions or a process explanation. Because these development methods are basic to technical and business communication, your instructor may assign them as separate exercises early in the course and ask that you apply these techniques to later report assignments as well.

UNIT 1.1 DEFINITIONS

Among the first things instructors do in a course is to introduce the basic concepts, methods, processes, and devices of their field. To make sure you understand the new material, instructors may provide formal definitions along with further explanations and illustrations. They are taking some pains to be precise because much later work will depend on a clear grasp of the basics.

Writing precise definitions of terms is also a good way to check your own understanding of the material. For this reason, instructors may assign definitions to you, or you may choose to write them yourself as a means of learning new concepts. Definitions of terms are also provided as parts of many technical, scientific, and business reports, particularly when these are written for general audiences.

POSSIBLE ASSIGNMENTS

Write a formal definition and then an extended definition of one of the following terms:

- ❑ Entrepreneurship
- ❑ Corporation
- ❑ Marketing research
- ❑ Opportunity cost
- ❑ Electrical resistance
- ❑ Optical fiber
- ❑ Modem
- ❑ Hypertext
- ❑ Program documentation
- ❑ Structured programming

Approximate length: one sentence for the formal definition, one to three paragraphs for the extended definition.

Explanation of a Definition

A *definition* is an explanation of something that sets it apart from all others. The explanation is primarily positive in that it specifies or describes essential features. But part of its task is also to set the limits of the term being defined—to say what something is *not.*

The best way to accomplish these ends is through a *formal* definition containing the following parts:

1. The term that names the thing or idea being defined
2. The class or category to which the term belongs (for instance, copper can be grouped with "metals")
3. The characteristics of the term that set it apart from others in its class (for instance, a metal that is ductile, malleable, a good conductor, and whose atomic weight is 63.54)

Any common abbreviations or acronyms for the term (for instance, Cu) may also be included.

These basic elements of a formal definition can be combined into a sentence or two. The following example has been numbered to show the basic elements:

> The *gross national product*① (GNP) is the total value ② of all the goods and services produced in a nation's economy in a given year.③

Formal definitions of terms are often used in reports at the point that technical terms or concepts are introduced. The ordering of the definition's parts may be varied to fit the context: "The total value of goods and services in a nation's economy, or its gross national product (GNP), is measured annually by the. . . . " Informal definitions, often given in parentheses, may also be used: "A nation's gross national product (total value of its goods and services) is measured annually by. . . . "

For complex or subtle concepts, the one-sentence formal definition, or its informal variants, may not be enough to help a reader or listener understand. In that case, the formal definition may become the first sentence of an *extended definition* of several paragraphs.

The additional clarification may be directed at parts of the explanation already given. In the gross national product case, for example, attention may be called to the fact that GNP is not goods and services but their value, and that such value would be stated in terms of the nation's currency.

Extended definitions may also offer illustrations (such as in Figure 1–1), examples, comparisons, and various informal techniques such as brief phrases, synonyms (*resistance* = *opposition*), or antonyms (*synchronous* vs. *asynchronous*). Informal techniques must not become so casual, however, that circular definition results: for example, "*Debugging* is when you fix *bugs* in a system."

If an extended definition is being given of a mechanism or process, it is helpful to include some breakdown of the parts to show their relationships and functions.

Examples of Extended Definitions

Here are two examples of extended definitions:

> ***Bit***
>
> A *bit* is the most basic unit of information in a computer system. In the computer, this information is conveyed by the presence or absence of voltage at a particular location. In other words, the computer recognizes only "on" or "off" conditions. Multiples of eight bits are commonly used to generate data or instructions.
>
> The bit is represented by the digits 1 or 0 in binary notation, and the binary number system is the basis of the computer's machine language. The binary code for the letter *c,* for example, is 1000011.
>
> The term *bit* is derived from *b*inary dig*it.*
>
> ***Market***
>
> A *market* is commercial activity created by people who are willing and able to buy the goods or services offered by sellers. Both buyers and sellers may act as individuals, or they may represent companies, organizations, and even governments.
>
> The willingness to buy results from the perception that a need would be met by the product or service, at a price acceptable to the buyer. The ability to buy, however, may be insufficient to create a market, as Western firms discovered in their initial ventures in the former Soviet Union.
>
> Willingness and ability to buy must in some cases be accompanied by authority to buy. Those who represent organizations or governments must be authorized to make the purchase decisions or a market will not be created.

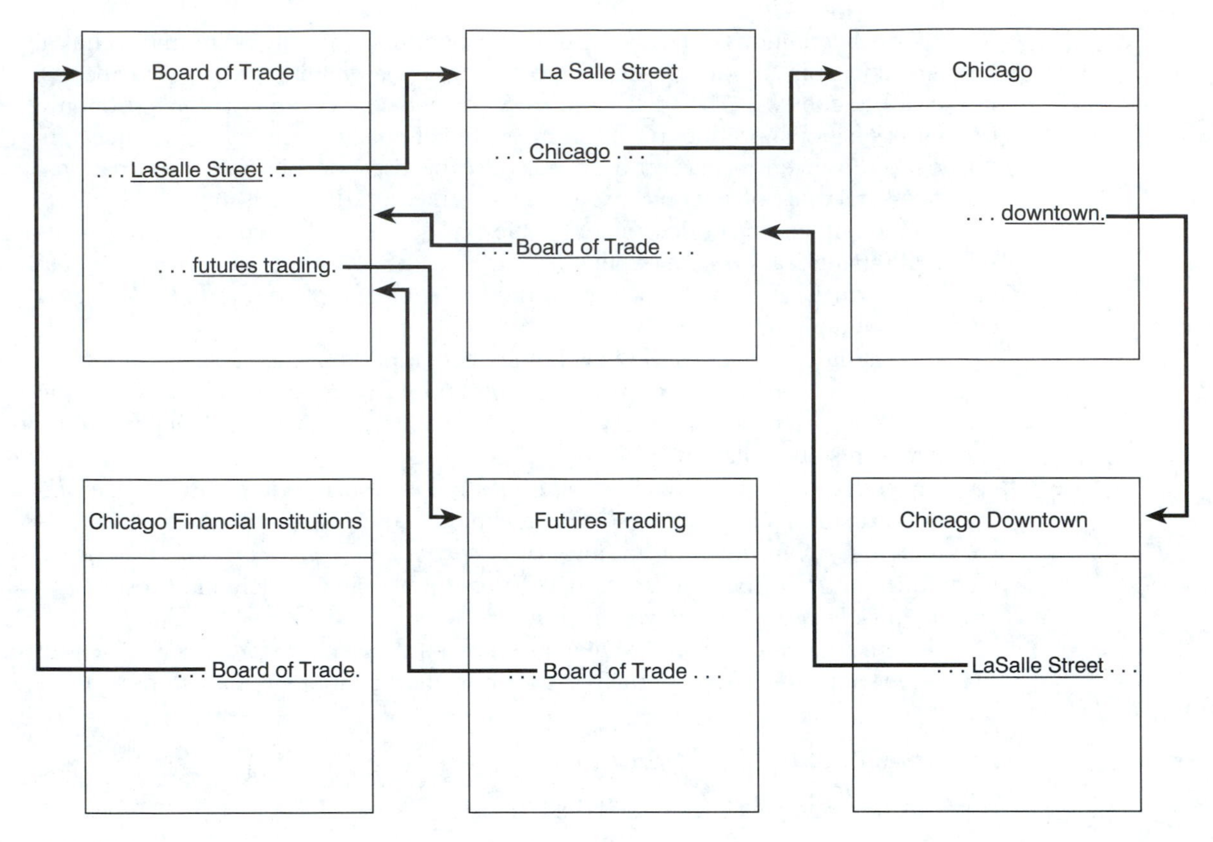

FIGURE 1–1
Illustrations may help clarify a definition. This schematic helps to explain the term *hypertext*.

UNIT 1.2 COMPARISONS

A comparison is a good learning exercise because it requires a close study of the things being compared. When you have to select criteria to compare things by, your attention becomes focused on important features such as cost or effectiveness for a particular purpose. Then too, having to finally decide which of several items is best for a purpose helps sharpen your judgment. In the final analysis, good judgment is the quality that education and experience are supposed to produce.

On the job, a comparison may be presented in an informal report that you are asked to prepare. Comparisons are also a common basis for feasibility reports.

POSSIBLE ASSIGNMENTS

Compare two or more types of one of the following:

- ❑ Word processing packages
- ❑ Laptop computers
- ❑ Documentation techniques
- ❑ Systems analysis methods
- ❑ Business organization methods
- ❑ Management styles
- ❑ Quality control methods
- ❑ Electronic mail programs
- ❑ Switches
- ❑ Control circuits
- ❑ Conducting materials

Approximate length: one to four paragraphs.

Explanation of a Comparison

In a *comparison,* your objective is to offer an opinion about which of two or more things is the best. You must also support that opinion by showing how the items being compared measure up against a set of criteria that you have selected.

Selecting criteria—the points to compare by—is the first and most important step in making a comparison. Accuracy, range of usefulness, price, reliability, appearance, and efficiency are examples of possible criteria. In most cases, you would choose a combination of such criteria, selected from those that logically relate to the purposes of your comparison.

A good way to select criteria is to focus on features that would make a difference if you were going to buy or use one of the things you're comparing. Putting yourself in the place of a consumer looking for value or a manager evaluating business practices, for instance, can help you choose criteria that suit the situation but reflect your own judgment.

Once you have selected criteria, you may find that each item in the comparison is best for certain applications. In that case, you can either discuss each application in turn, or you may want to direct your comparison toward a single application.

For instance, if you compared two types of athletic shoes, you might conclude that both are durable and reasonably priced but that one is better for tennis and the other for running. You can write the comparison to show that each shoe is better for a different application, or you can write it to show that one is a better tennis shoe—or running shoe (see Figure 1–2).

Remember also that some criteria are more important than others. If items don't compare closely on these important criteria, there is no point in drawing out the comparison on less important criteria. In other words, don't compare two automobiles on roominess or comfort unless they are first comparable on such criteria as reliability and cost.

A COMPARISON OF ATHLETIC SHOES

	Cost	Durability	Applications
Shoe A	Medium	High	Good for tennis
Shoe B	Medium	High	Good for running

A COMPARISON OF RUNNING SHOES

	Cost	Durability	Heel-to-toe action	Arch support
Shoe A	Medium	High	Poor	Fair
Shoe B	Medium	High	Good	Good

A COMPARISON OF TENNIS SHOES

	Cost	Durability	Side-to-side action	Grip on smooth surface
Shoe A	Medium	High	Good	Good
Shoe B	Medium	High	Poor	Fair

FIGURE 1–2
Different approaches to comparing athletic shoes.

Once you have selected the criteria and clarified the applications of the items being compared, you can more objectively decide which item is best. At this point you are ready to report on your decision.

Start by reviewing the planning and development process discussed in the Introduction. In this case, you will already have done several of the planning steps, and you can move quickly to a decision on format, develop a working outline, and then begin to write.

A good general format for a comparison report includes the following sections, in order:

1. State the purpose of the report: a comparison of several items for certain applications.
2. List the criteria you have selected.
3. Detail how each item rates against the criteria. (A chart may suffice.)
4. Discuss the ratings in relation to the applications you have chosen; give your opinion on which item is best.

After a first draft is completed, leave some time for revision. If your comparison is directed to mechanisms or processes, for example, you may find that adding graphics of a visual or schematic kind can highlight differences.

Example of a Comparison

Figure 1–3 gives a comparison of silver, copper, and aluminum for wiring applications.

UNIT 1.3 SUMMARIES

In a fast-paced world, good summaries of information are essential. In your own career, skills that are developed by summary exercises—like being able to distinguish essentials from nonessentials—will help you manage the volume of reading you have to do and the reports or memos you have to write.

In business and industry, a long formal report is often preceded by an *executive summary* for the busy reader who lacks time to read the full report. An executive summary is equivalent to the *informative abstract,* which describes a report's findings and conclusions, rather than the *descriptive abstract,* which outlines the organization of the report. Although the descriptive abstract has its uses as a guide

SILVER, COPPER, AND ALUMINUM IN WIRING APPLICATIONS

This paper compares silver, copper, and aluminum to determine which is best for electrical wiring. The criteria used are cost, conductivity, and compatibility with existing electrical systems and wiring practices. Each metal is rated on these criteria in the following chart.

Material	Cost per Cubic Inch	Conductivity (Using Copper as a Reference: Cu = 1)	Compatibility Problems
Silver	$41.60	1.05	No
Copper	$00.20	1.00	No
Aluminum	$00.08	0.61	Yes

An important fact in this comparison is that conductivity is increased by using thicker wire. Therefore, a thicker aluminum wire can conduct as well as, or better than, a thinner copper or silver wire. According to the chart, aluminum wire twice as thick as copper or silver wire would give the best conductivity (1.22) at the least cost ($0.16). In fact, aluminum is used whenever possible, especially by the power companies. Many high-voltage wires are made of aluminum.

However, aluminum corrodes easily when it comes into contact with copper, brass, steel, or bronze, and it is difficult to solder. Thus, aluminum would pose problems in wiring homes where connectors are frequently made of these other metals, or in constructing circuits in the lab. In these applications, copper is the best choice because of its cost advantage over silver. Silver is used only when its slightly better conductivity is very important, as in receivers that must pick up faint radio frequency signals.

FIGURE 1–3
Example of a comparison.

to particular information in a long document, this unit focuses on the informative purpose—on basic techniques of summarizing information and ideas.

POSSIBLE ASSIGNMENTS

- ❑ Summarize the decisions from a committee meeting.
- ❑ Summarize the findings from a series of related lab assignments.
- ❑ Summarize a newspaper or magazine article selected by the instructor.
- ❑ Summarize a class lecture or a presentation by a speaker.
- ❑ Summarize all or part of a reading assignment.
- ❑ Provide an informative abstract (or executive summary) for a longer report.

Approximate length: two to three paragraphs.

How to Summarize

The general idea in summarizing is to "say as much as the original but in fewer words." This can be achieved by including every essential point while excluding details as well as anything that is not in the original material, especially personal opinions or knowledge. A good summary emphasizes the same points as the original, leaves out most details and examples, and quotes sparingly.

We summarize quite naturally in ordinary conversation. When someone asks, "What happened?" we instinctively relate the main points needed to understand the event, and we compress the sequence of actions: "I drove to the Bijou. I parked the car. I bought the ticket and entered the theater. I found a seat, etc." becomes "I saw a movie last night." This natural tendency to compress events can be applied while you analyze the material you wish to summarize.

The first question to ask is, "What is this about?" In written material, such as an article, you may need to look no further than the title to determine the subject. If the title is unhelpful, check the introductory sections of the article, where the subject is usually identified. Then ask, "What does the author say about this subject?" Most articles are written to express a thesis, that is, a conclusion or finding about the subject. The thesis is basically the author's reason for writing. In other words, articles (a) have a subject and (b) say something about it. By determining the subject and the author's thesis, you can track the points you need to summarize.

Sometimes you may need to ask an additional question: Granted that the subject is X, just what *is* X anyway? Even if you know what X is, you may need to briefly define it for your readers.

Take notes as you read the material, and try to answer these key questions first. In the case of the article about e-mail summarized in this unit, the process worked roughly as shown in Figure 1–5.

Once you have a good overall idea of the author's subject and purposes, take notes on the major facts and details of the article. When you're finished, go back and cross out most examples and many details of the kind that add nothing new to the author's thesis. Such details may support the thesis but do not extend it into new areas. Such details should not be a part of your summary. Also, cross out any personal opinions or comments you have added in taking notes.

To write the summary itself, organize your notes for an orderly approach. A good general approach for writing an article summary is the following:

1. Name the source of the material being summarized—for example, title, author, name of publication, and date.
2. State the subject.
3. Define the subject if necessary.
4. State the author's thesis.
5. Elaborate or explain the thesis.

How much elaboration of the thesis you should do can be guided by the purposes of your summary. As a rule, you want readers to understand the points being made, but you do not have to *prove* the thesis. That's the author's job.

Example of a Summary

Figure 1–4 is a reprint of an article from the *Wall Street Journal,* February 1, 1993 (reprinted with permission). Fig. 1–5 shows a set of notes taken while reading the article. Compare these documents to the article summary (Figure 1–6) that follows.

UNIT 1.4 INFORMAL TECHNICAL DESCRIPTIONS

Imagine being on the phone and trying to order a part for which you have no name or part number. You'd need to describe the item so the other person could get a good mental picture of it. Writing technical descriptions improves your skill at describing things in words only. This is a useful skill in many work situations. Dealing with customers or co-workers often requires brief, efficient description of this kind, in both writing and speech.

POSSIBLE ASSIGNMENTS

Briefly describe one of the following:

- ❑ Simple hand tool
- ❑ Floppy disk
- ❑ Type of visual aid
- ❑ Menu screen
- ❑ Battery

E-MAIL DELIVERS FIRST-CLASS PACKAGE OF USEFUL SIMPLICITY

Walter S. Mossberg

Electronic Mail is an increasingly important form of business and even personal communication. But it's way too complex and costly for average computer users, right?

Wrong. Just about anybody with a personal computer and a modem—the device that hooks computers up to the phone system—can subscribe to a commercial e-mail network for reasonable prices. Then you can use a variety of software, including some simple programs, to exchange messages from your screen with millions of people in the U.S. and abroad.

Both American Telephone & Telegraph and MCI Communications, the titans of long distance phone service, also offer commercial e-mail services, called AT&T Mail and MCI Mail. Another popular service is run by CompuServe, the big on-line database service owned by H&R Block.

In most cases, you simply bang out messages to send, and read messages you have received, while "off-line"—that is, when you're not connected to the system. Then, at the touch of a key or the click of a mouse, the software will automatically dial up your service, rapidly send all the messages you've composed since your last call, and just as quickly pull into your own computer any messages others have left for you.

Each user has an "electronic mailbox," a small part of the disk storage space on the systems' mainframe computers. Each mailbox has a distinct address, plus a separate password for security. Users can find addresses in on-line listings.

These commercial e-mail systems even allow you to send messages to people who lack computers. For an extra fee, you can direct that any message be delivered as a fax or letter. For another additional fee, some services will print faxes and letters on copies of your own letterhead with copies of your signature.

Most of the big services also permit members to send and receive messages to and from account holders on competing networks, and to people who belong to in-house e-mail systems at large companies that have configured their networks to connect to the outside world. In most cases, you can also link up with people who use the Internet, a vast interlocking system of computer networks mostly at universities and government agencies.

This adds up to a tremendous tool for individuals and small businesses, a way to help even the playing field with the big boys. Sitting in a home or storefront office, with any mainstream brand of computer, an e-mail subscriber can compose a single message and, with a single keystroke, fire it off to dozens of suppliers or customers. One minute she can get the same instant price updates long

FIGURE 1–4
An article from the *Wall Street Journal,* February 1, 1993 (pp. 42–43).

- ❑ Automobile headlight
- ❑ Face of a meter or scope
- ❑ Keyboard layout
- ❑ Office layout
- ❑ City square

Approximate length: one to three paragraphs.

available to big companies, and the next minute she can send an e-mail note to her daughter at college.

Because the systems hold mail for members until they call in, they avoid the problem of "telephone tag." And since the computers transmit messages quickly, you can cram much more information into a brief e-mail call than into a regular voice message. Commercial e-mail rarely swamps users with floods of messages, unlike in-house corporate e-mail systems, where messages are free and don't require placing a phone call.

So how much does it cost? In most cases, you can reach the commercial e-mail systems by placing a local call or dialing a toll-free number. Users are typically charged only for messages they send, not those they receive. The cost is based on the length of the messages, not the time spent on-line. For long documents, e-mail can get costly. For small letters, it can be quicker and cheaper than overnight delivery services.

Pricing plans vary. They can be complex. The basic charge for my favorite service, MCI Mail, is $35 a year. Each message costs 50 cents for the first 500 characters, or roughly 75 words. The second 500 characters, and each of the next several 1,000 characters, cost a dime. The rate drops to a nickel per thousand characters beyond 10,000 characters. A message as long as this column would cost about $1 to send through MCI Mail to anybody in the U.S. or abroad.

Software is a crucial issue. It's best to use special programs called "front ends," which streamline and automate the process of writing messages, dialing into your service, and reading and storing your e-mail. MCI and AT&T both offer their own front-end software, but several other programs are better, in my view.

One is the Wire, from Swfte International, a $90 front end for MCI Mail customers with IBM-compatible PCs running Microsoft Windows. CompuServe's $30 membership kit includes a good program, Information Manager, in versions for DOS, Windows and Apple Computer's Macintosh. Software Ventures' $140 Microphone program for the Mac and Windows also includes good front ends for MCI and CompuServe.

It would be ideal to have one piece of dedicated, simple software that can automatically manage your mail on multiple services. One Button Mail from Sigea Systems, Weston, Mass., an IBM-compatible program that handles both MCI and CompuServe, is great but costly at $250. Hewlett-Packard plans to introduce this spring a $99 Windows front end that can handle four major e-mail services.

One final advantage of e-mail is worth mentioning. I have yet to receive an electronic junk message declaring: "YOU MAY HAVE WON A MILLION DOLLARS." But if some sweepstakes company reading this wants to give away a lot of money via e-mail, I can be found on MCI Mail at the address "WMOSSBERG."

Explanation of an Informal Technical Description

A technical description tells about its subject in detail. It draws a verbal picture. It answers questions such as "What is this?" "What does it look like?" and "What is it used for?"

Unlike a definition, which seeks to categorize its subject, a description tries to help a reader visualize it. Descriptions emphasize physical and spatial characteristics, while definitions of the same objects might focus on functions. A definition of a "city square," for example, might stress its role in providing open space for gatherings and recreation. A description of a square would likely cite its dimensions, location, surrounding buildings, and internal physical features.

Subject

Electronic mail (e-mail)

- Actually subject is the reasonable cost and effectiveness of e-mail services

What is it?

Not a technical definition, but:

- A form of business/personal communication
- Messages on your computer sent and received via modem that links you to phone lines

How does it work?

- Type out messages while off-line
- Use software to send your messages and pull in messages from others
- Each user has a "mailbox"—a distinct part of the mainframe memory accessed by a password for security

Capabilities

- Can send to those who don't have computers
- Send to competing e-mail services
- Can link up to Internet, etc.

Advantages

- Helps small businesses compete—get info on prices, promote to customers
- Avoids telephone tag by holding messages

Costs

Quite reasonable—charge only for messages you send

- Based on length of message
- MCI Mail = $35 a year
  $.50 first 75 words
  .10 next 75 words
  .10 next 150 words, etc.
- His article (@900 words) would cost $1.00

Software—a key

- For sending & receiving messages—managing your e-mail
- "Front-end" software does this best
- Costs are $90–$140, up to $250

FIGURE 1–5
Sample summary notes.

ARTICLE SUMMARY

"E-Mail Delivers First-Class Package of Useful Simplicity," by Walter S. Mossberg, in his "Personal Technology" column in the February 1, 1993, *Wall Street Journal,* makes a case for the advantages of electronic mail as a form of business and personal communication. The author argues that e-mail, as provided by commercial services, is a cost-effective alternative to phone, fax, or mail.

Costs of messages sent are about $.50 for the first 75 words, and $1.00 for about 900 words. An annual fee may be $35, and software for managing the flow of messages costs $90–$150. A computer and a modem are also necessary to send and receive messages.

For the cost, e-mail subscribers can type out messages while they are off-line and then use the management software to both send their messages and pull in messages from others. The efficiency and reasonable cost of e-mail evens the field for small business owners, who can get the same information in a timely manner as the large firms, and who can quickly send messages to suppliers and customers.

FIGURE 1–6
Summary of the *Wall Street Journal* article.

A technical description is *informal* because it follows no set pattern of development, not because it is imprecise. The description should use precise words and measurements, such as "189 pounds," rather than "heavy," or "2 1/2 inches high," instead of "small." A reader familiar with an object being described should be able to recognize it from the description; someone who is unfamiliar with the object should be able to picture it after reading the description. The reader should also be able to understand what the object does or what characteristics it has.

While there is no expected format for an informal technical description, a variety of methods can be used to evoke a mental picture of the subject: giving measurements and shapes, comparing the object to a more familiar one, explaining the object's parts in a logical order, and so on. A common method is to move from the general to the specific:

1. Give a brief statement of the object's functions.
2. Give an overall physical description of the object.
3. Describe the object's parts in a consistent order.

How much description of the parts should you provide? As in most writing situations, the answer depends on the way your description will be used. Because the informal description is generally not intended as a guide for maintenance or other technical applications, the amount of detail is limited. Emphasis is placed on the general aspects of size, shape, and function.

Examples of Informal Technical Descriptions

Two examples of informal technical descriptions are provided in Figures 1–7 and 1–8.

THE PIE CHART

A pie chart is a geometric figure based on a circle divided into sectors. The sectors are formed by lines drawn from the center of the circle to its edge, much as a pie is cut into pieces. From two to eight labeled sectors are common. The sectors can be "raised" or shaded or colored for added contrast.

The pie chart is a visually effective way to compare parts to each other and to a whole. For example, a circle with one very small sector labeled "Female Students" and the remaining sector labeled "Male Students" would show a college's uneven student population at a glance.

A pie chart with exact dimensions will have sectors directly proportional to the percentages they represent. For example, if 15% of the students are female, the chart would include a sector with an angle of 54°, which is 15% of the 360° in the circle.

FIGURE 1–7
An informal technical description.

UNIT 1.5 FORMAL TECHNICAL DESCRIPTIONS

The techniques of formal technical description may be used to convey a clear picture and a good understanding of a mechanism or device. Formal, rather than informal, description is more apt to be used when the information is needed by specialists or product users concerned with installation, functioning, or maintenance.

THE INCANDESCENT LAMP

The common light bulb, which has a coiled tungsten filament sealed in a glass bulb, is an incandescent lamp. Such lamps produce light when the filament is heated to about 2500° to 3000° C by an electric current passing through it.

The coiled filament is suspended between a pair of lead-in wires projecting from a stem within the bulb. The bulb is filled with an inert gas to retard wearing out of the filament. The base of the bulb is threaded on the outside to allow the unit to be screwed into a receptacle providing electric current at the flick of a switch.

Light bulbs are commonly frosted to diffuse the light and provide more even illumination.

FIGURE 1–8
An informal technical description.

POSSIBLE ASSIGNMENTS

Write a formal technical description of one of the following:

- ❑ A high-resolution laser printer to a business executive
- ❑ A private branch exchange (PBX) system to a facilities manager of a corporation
- ❑ A particular model of an oscilloscope to a service technician
- ❑ A computer workstation to a businessperson working from home
- ❑ A microprocessor trainer to an electronics instructor
- ❑ A modem to a first-term college student
- ❑ An automobile carburetor to a high school student
- ❑ A disk drive to a college freshman

Approximate length: one to five pages, depending on the purpose of the description.

Explanation of a Formal Technical Description

A formal technical description rarely stands alone. It is usually part of a longer document such as a formal report, user's manual, or service manual. In any case, it has a distinct purpose: to help the reader picture the object being described and understand its functioning in some detail.

The reader may need this understanding for various reasons. The description may help orient the technician before service or repair activity. The description may help the nontechnical reader who needs to use the object. Or it may help a reader understand the object well enough to judge its practicability, as in the case of an operations manager debating whether her company should manufacture a newly designed device.

The techniques of effective description include suiting your language and selection of detail to the reader's background. You should also seek to describe objects without *relying* on the use of illustrations. A picture may be worth a thousand words, but the words should guide the reader through the visual exploration, not leaving any observations to chance. Illustrations should clarify verbal description; the verbal description should be complete and able to stand alone if the illustrations were removed.

A good approach to a formal technical description is to first provide a *definition* of the object linked to an explanation of its *function.* But before you begin, decide how much your audience already knows. Are you writing to a technician who is familiar with multimeters, for instance, and just needs to know the particulars of the model you're describing? Or are you writing to a business executive who may not even know what functions a multimeter serves? In the first case, you would only need to explain what distinguishes this meter from others ("The KDM-385A is a compact battery-powered digital multimeter that . . . "); for the executive you'd have to broaden your definition to include "multimeter" itself.

Knowledge of the audience can provide an effective guide for your description. This awareness can help you determine, among other things, how much detail to include, how technical your language should be, and what kinds of illustrations to provide.

After an introduction to the object and its functions, go on to *describe* the overall appearance of the object. Consider the factors involved in visualizing it: shape, size, weight, materials used, spatial relationships, and the like. Build a mental picture yourself, using only the details you provide and check whether you "see" the complete object. After writing the description, read it to someone who is unfamiliar with the object. Did the mental picture match the reality? Refine your description according to the feedback you receive.

To deepen the reader's understanding of the object, break it down into a list of component *parts* next. These may be parts that go into actual assembly of the object, such as the engine, transmission, and frame of a car; they may also be materials the object is composed of, such as the inner and outer materials of a resistor or a silicon chip. In the case of mechanisms or equipment with many parts, list major sections that you will describe further later.

Treat each of these major parts separately. *Describe* each one, breaking it down even further if your readers require it, and *explain* how each one contributes to the object's function or operation.

By providing these breakdown levels, you are inviting your reader to "take a look inside"—sometimes at features that would not be visible in looking at the object itself. This part of the description should be presented carefully and systematically so the reader does not lose sight of the general picture in the levels of detail.

Following is a suggested general approach for writing the formal technical description:

1. Definition (in terms of function)
2. Overall description (dimensions and appearance)
3. A list of component parts
4. Description of parts
5. Relation of parts to the whole

To apply this format in a document of some length, you may find it useful to review the development strategies suggested in the Introduction by referring to the checklist on pages 29–31.

Example of a Formal Technical Description

The description in Figure 1–9 was written as the introduction to a user's manual.

THE HP LASERJET 4 PRINTER

The Hewlett-Packard HP LaserJet 4 model C 2001A is a versatile professional printer that delivers excellent print quality at eight pages per minute on various paper sizes and on envelopes, labels, and transparencies. The LaserJet 4 offers 600 dpi (dots per inch) resolution and 45 internal scalable typefaces. It occupies about the same desk space as a personal computer.

Electrical requirements for operation are 100 to 115 volts, or 220 to 240 if necessary, and only basic operational maintenance is required. The electrophotographic cartridge contains toner, or dry ink, and is completely disposable and replaceable. No pouring of toner or threading of ribbons is required.

The LaserJet 4's height is 10.2 in at the front and 11.7 in at the back, as the unit slopes up to accommodate print output (see Figure 1). The unit is 16.4 in wide and 15.9 in deep. An additional 4 in of space should be allotted behind and to the sides of the unit to avoid buildup of heat from operations.

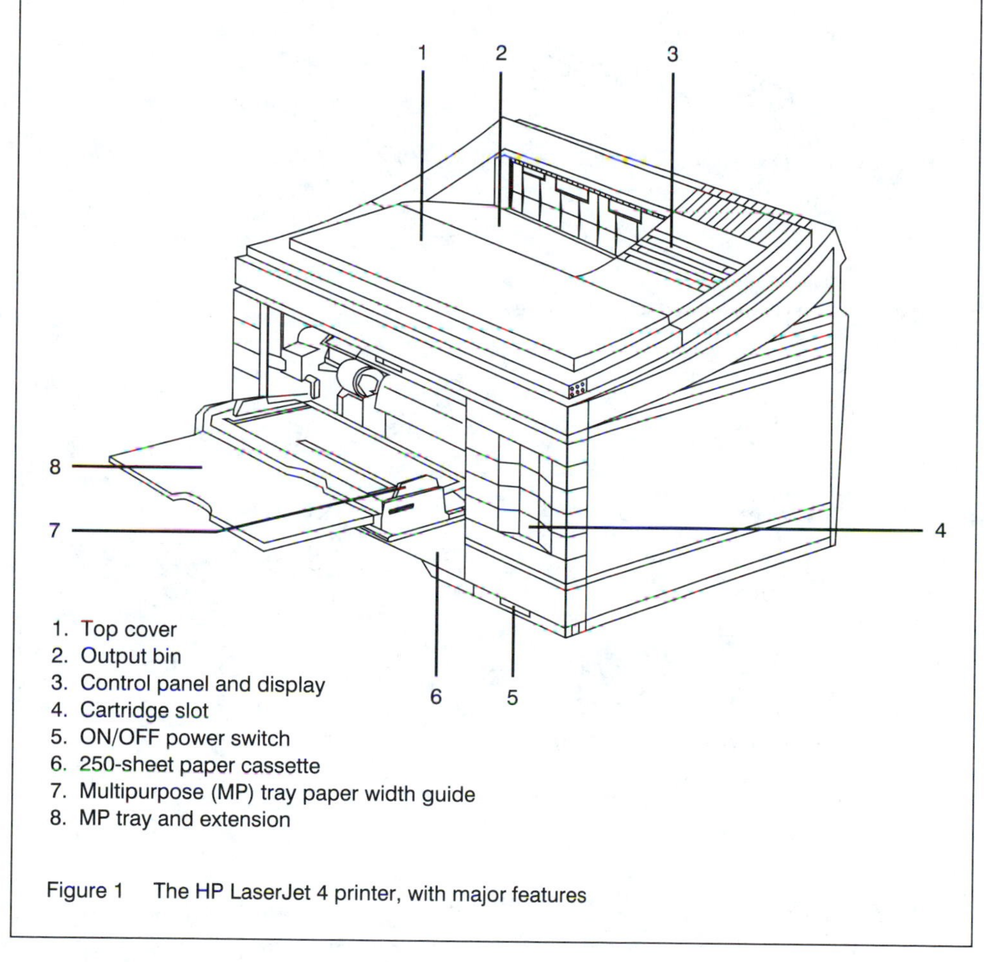

Figure 1 The HP LaserJet 4 printer, with major features

FIGURE 1–9
A formal technical description (pp. 49–51).

An output slot 9.5 in wide guides printed copies to the top of the unit, in the space next to the control panel. A slot for a 250-sheet paper cassette occupies the bottom of the front panel, while a multipurpose paper tray sits above the cassette. To the right of the paper trays is a 3/4 by 4 in cartridge slot for additional fonts and typefaces.

The toner cartridge for the printer lasts about 6,000 pages at a medium print density setting. Toner cartridges are approximately 8.2 in wide, 4.7 in deep, and 3 in high at the back face of the cartridge (see Figure 2). Cartridges are removed and replaced by sliding them along guides into the space under the top cover of the printer (see Figure 3). A cartridge fits snugly into its space when it is pushed forward and back as far as it will go into the unit.

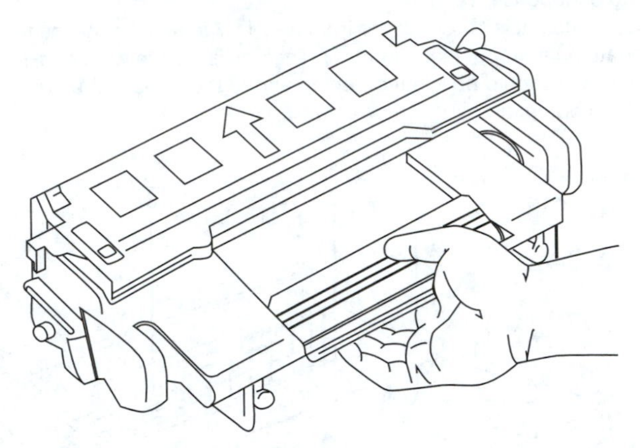

Figure 2 The HP LaserJet 4 toner cartridge

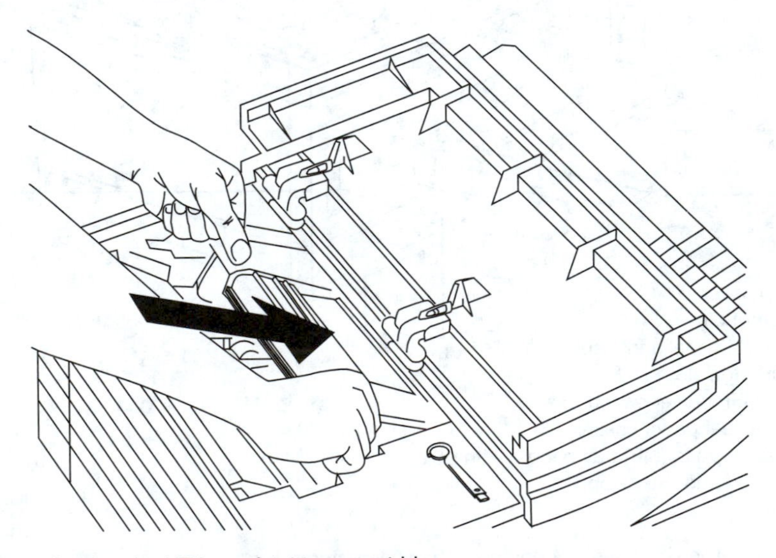

Figure 3 Seating the HP LaserJet 4 toner cartridge

FIGURE 1–9, *continued*

In typical operation, 8-1/2 by 11 in paper is loaded into the 250-sheet tray so that the stack rests under the metal holders at the corners of the tray (see Figure 4). The tray is then pushed back into the slot at the bottom of the unit (Figure 5), and printing can commence.

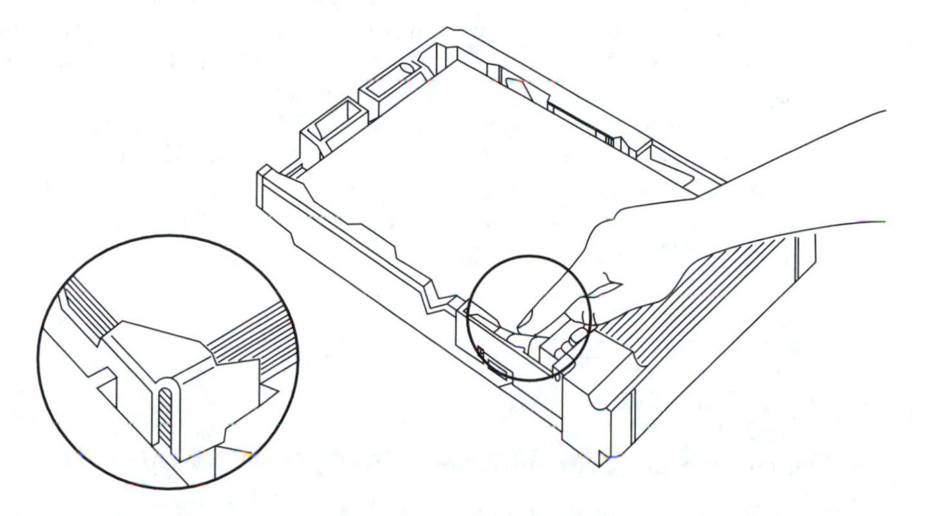

Figure 4 Loading the HP LaserJet 4 paper tray

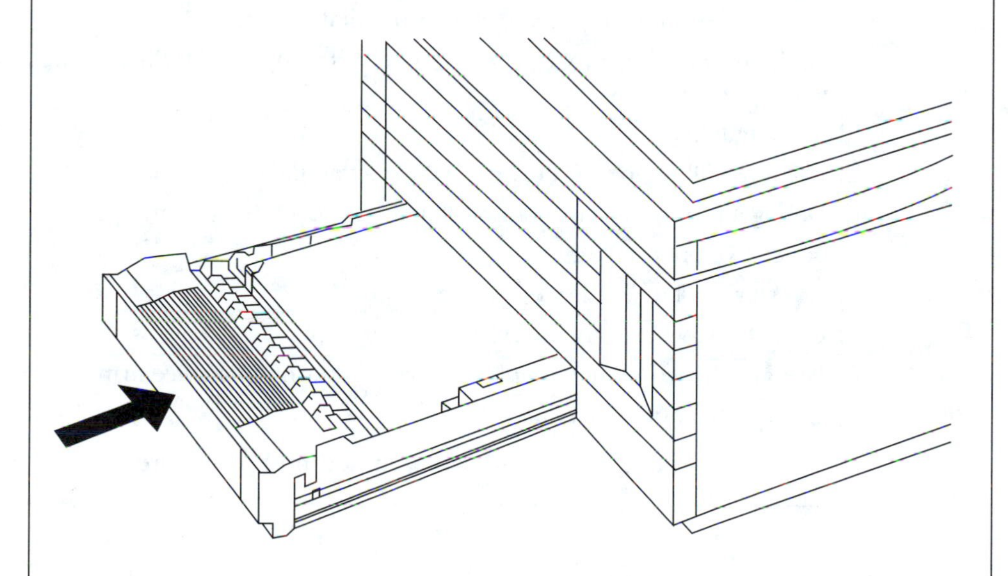

Figure 5 Inserting the 250-sheet paper tray in the HP Laser Jet 4 printer

UNIT 1.6 INSTRUCTIONS

The ability to give precise instructions is a valuable professional skill. You can create good customer relations for your company if you can help customers understand how to use your products. Within the company, this skill is the key to effective employee training and supervision. It becomes critically important in writing about procedures that could lead to injury or damage if done improperly.

If you have ever struggled to follow a confusing set of directions to a strange part of town, you know that giving clear instructions is a much needed personal skill as well. And as every wage earner knows, the instructions to filling out an IRS return are every bit as "taxing" as the tax tables.

POSSIBLE ASSIGNMENTS

Give written or spoken instructions for one of the following procedures:

- ❑ Boot up a personal computer
- ❑ Send an e-mail message
- ❑ Use a spreadsheet to prepare a departmental budget
- ❑ Perform a mathematical operation, such as solving a set of linear equations with three variables
- ❑ Plot a graph
- ❑ Use an oscilloscope to verify the gain of an amplifier
- ❑ Debug a program, or troubleshoot a microprocessor system
- ❑ Splice two optical fibers
- ❑ Transfer a call to another extension
- ❑ Conduct a focus group interview
- ❑ Prepare an income statement for a manufacturing or service firm
- ❑ Chair a committee meeting

Approximate length: one to two pages for most procedures.

How to Give Instructions

Instructions may be part of a user's manual, a lab manual, or a maintenance guide, and they are a familiar part of brochures accompanying consumer products. What they try to do is simple in concept, if not always in execution.

When you give instructions, you tell someone exactly what to do, in sequence, to carry out some physical process. The instructions are broken down into a sequence of manageable steps, with each step following in logical succession. An introduction and helpful explanations or cautions may be included, but these should be kept separate from the steps themselves. These supporting comments should consist of only the information necessary for the listener or reader to understand the requirements of the task.

For instance, if you wanted to tell a child how to plug in a lamp, you probably wouldn't explain the theory of electricity first. You would, however, caution her not to touch the prongs or the socket. Also, for the child's sake, you would give this caution *before* the instructions and give it as much emphasis as possible.

Among the kinds of information you might include in the introduction are the following:

1. Warnings, notes, and cautions. These should be prominently displayed and emphasized (for example, capitalized or underlined).
2. Any special tools or equipment or materials required to carry out the instructions.
3. The conditions under which the procedure should or should not be carried out: For example, never load film into the camera in direct sunlight.
4. Illustrations that support explanation of complex steps in the procedure.
5. A definition of the procedure or an explanation of its purpose. If the procedure is familiar (changing a flat tire), nothing need be said. If the procedure has a more subtle purpose (that of a marketing research study, for example), explain its value or intent.

Whereas a process explanation (see Unit 1.7) tries to convey a general understanding of a procedure, a set of instructions is concerned with helping someone to perform it. The introduction, therefore, is typically brief.

In the instructions themselves, the emphasis is on (1) maintaining a proper sequence and (2) dividing the process into sufficiently small steps to be easily followed. To write clear instructions, go over the procedure first—either in your mind or in actual performance—while you take notes. This method will also help you put the instructions into the proper sequence.

Instructions are given in the form of commands or directions: "Trim the end of the longer wire" or "Insert the probe into Port B." Each step of written instructions should be numbered, and each new activity of a procedure should be a separate step. If you find yourself saying "Push this *and* pull that," check whether you actually should have two separate steps. Some steps do occur simultaneously, as in "Hold the red button down *while* you move the control lever forward." Steps may also be combined if they are continuations of each other: "Move the pointer to 'misc.books.technical' and click twice."

Generally, steps should be kept as short and simple as possible. An average sentence length of 10 to 15 words should be maintained so steps can be understood and followed easily.

You can test your written instructions for user friendliness by following them *exactly* yourself, or you can have someone else follow your instructions while you watch. If there are problems, check whether steps have been separated properly and are in the right sequence. Remember also that physical processes must be described precisely. Use the methods of good technical description: Focus on shape, size, color, position, and other indicators that help the reader distinguish among objects.

Example of a Set of Instructions

Figure 1–10 provides a set of instructions for setting up an e-mail program.

HOW TO SET UP FOR E-MAIL WITH EUDORA

Sending electronic mail across the Internet requires an applications program such as Pine, Elm, Eudora, or a similar version. This software helps a user format and edit a message and then sends it to the recipient's mailbox. Eudora is often grouped with other applications in such programs as Plug 'n' Play.

NOTE: *These instructions apply to a PC user in a Windows environment running Plug 'n' Play applications.*

1. Use the attached mouse to position the pointer on the computer screen over the Plug 'n' Play Internet icon.
2. Click the mouse button twice rapidly to bring up the Plug 'n' Play Internet applications screen (see Figure 1).

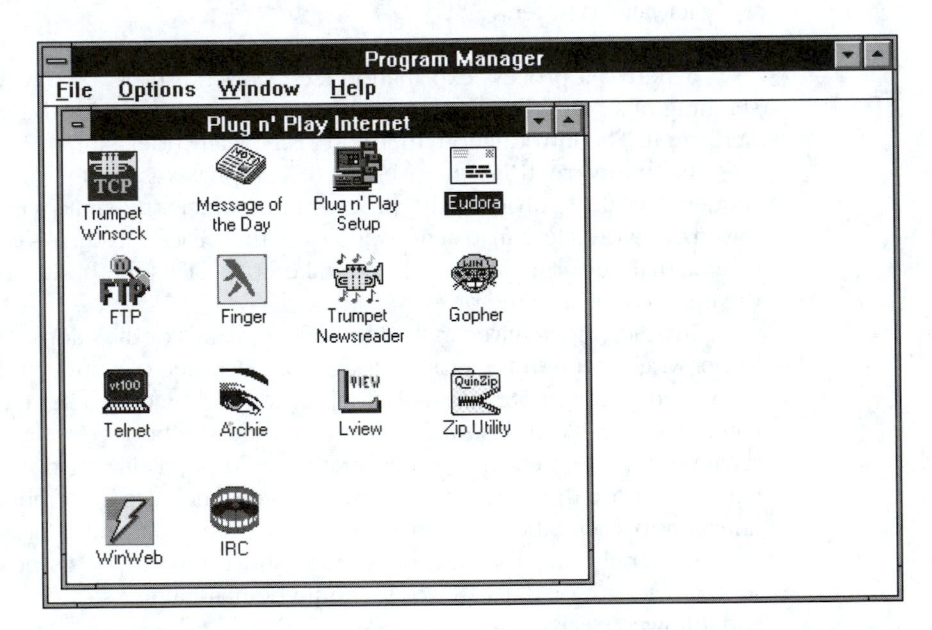

Figure 1 A Plug 'n' Play Internet screen showing the applications available in the program

FIGURE 1–10
A set of instructions (pp. 54-55).

3. Using the mouse, position the pointer over the Trumpet Winsock icon on the screen.
4. Click twice rapidly.
5. Wait for the Trumpet Winsock program to request your login password.
6. Use the keyboard to type in your password, and enter it using the Enter key.

NOTE: *For security reasons, the password will not appear on the screen as you type, but will be represented by asterisks equal to the number of letters in the password.*

7. Wait for the password to be accepted and the "PPP Enabled" message to appear on the screen.
8. Move the pointer to the upper right corner of the screen and click once on the square containing the downward arrow.

CAUTION: *Do not proceed unless the Trumpet Winsock icon is shown outside the Plug 'n' Play Internet window, indicating the former has been "minimized".*

9. Position the pointer over the Eudora icon and click twice.
10. Wait for the program to request your password.
11. Enter your password as in Step 6.
12. Wait for your e-mail messages to be logged in.
13. Acknowledge that "you have new mail" by clicking once on the "OK" window.
14. Read your mail or send a message with the help of Eudora.

UNIT 1.7 PROCESS EXPLANATION

Processes are basic to technology, business, and work of all kinds. Particular processes must be clearly understood so that they can be evaluated for efficiency and cost-effectiveness. By learning a successful approach to describing a process, you acquire a tool that can be used in many situations on the job. You may, for instance, need to explain a process as part of an orientation for customers or other employees. You may also need to use the technique in longer documents such as lab reports, proposals, and user's manuals.

POSSIBLE ASSIGNMENTS

Explain one of the following, either in writing or speech:

- ❑ How a disk drive reads a hard disk
- ❑ The operation of a sequential logic circuit
- ❑ An accounting cycle
- ❑ How to troubleshoot a power supply
- ❑ How to hook up and operate a logic analyzer
- ❑ A production scheduling or control technique
- ❑ A system life cycle
- ❑ The interaction of water sources in an ecological system
- ❑ How to design a program using top-down structured programming techniques
- ❑ A collaborative problem-solving approach

Approximate length: a few paragraphs to a few pages.

How to Explain a Process

A set of instructions directs a person, step by step, in performing a process, regardless of whether she understands its broader purpose. The language of instructions is in the form of commands: "Turn this" or "push that." An explanation of a process, however, strives to make the reader *understand* the process—how, when, where, and why it is performed—even if he never performs it or witnesses it personally. And some processes that must be explained are performed by mechanisms or occur naturally. So instead of instructing, you *describe.*

As a first step in your process explanation, analyze the process thoroughly so you understand it yourself. Then set up your explanation in three sections: (1) an introduction, (2) an overview of the major stages in the process, and (3) a detailed breakdown of each stage.

1. The introduction should provide some background for the reader. It may include the conditions necessary for the process to occur, the equipment or materials needed to perform the process, or why the process occurs. In selecting information of this kind, you should include the points that helped *you* understand the process. You may also use more or less detail, depending on how much your reader already knows.

2. The overview of the major stages in the process should identify no more than six stages, given in the order in which they occur. As an overview, it should avoid details and complications. Its purpose is to help the reader see the process as a logical whole. The overview also helps the reader follow the third part of your explanation.

3. Breaking down each major stage into its details will be the largest part of your process explanation. This section should guide the reader smoothly through the series of actions or occurrences that make up the process. To help the reader understand, explain not only what is done but how and why. Illustrations can be used to support the explanation.

If actions occur simultaneously in the process, alert the reader to this condition. Then explain each action in turn.

An outline of a three-part process explanation follows:

I. Introduction (background for understanding)
II. Overview of major stages (no more than six)
III. Detailed explanation of stages
  A. Major stage A
    1. Action 1
    2. Action 2
    3. Action 3
  B. Major stage B
    1. Action 1
    2. Action 2
    (and so on)

To help you plan, develop, and revise your explanation, review the steps suggested in the checklist on pages 29–31.

Examples of the Process Explanation

Two examples of process explanation are provided (see Figure 1–11).

UNIT 1.8 FIGURES AND TABLES

This unit on figures and tables is for reference only. No "typical assignments" or length specifications are given because the materials discussed here are almost never used in freestanding assignments. Rather, figures and tables are common illustrative devices included within writing projects and reports.

The general terms *graphics* and *visuals* refer to a wide variety of illustrative materials used to support the words in a written report. The two major classes of visuals are *tables* and *figures.*

Tables are groupings of data according to categories so that quick comparisons and interpretations can be made. Figures include various illustrations or visual explanations. Actual photographs or drawings may be included as figures. More common kinds of figures include diagrams (such as schematics), graphs (such as line graphs and curves), and charts (bar charts, pie charts, flowcharts, organizational charts, and others).

CHECK PROCESSING

Approximately 90% of the dollar volume of business transactions in the United States every day is handled by check. For example, bills are paid by check, banks loan money by check, and paychecks are deposited to cover personal checks that will be written later. The use of checks increases safety and convenience, provides a record of payment, and helps the Federal Reserve System regulate the money supply.

Your check authorizes your bank to deduct from your account the amount of money you specify and tells it to pay the person or organization you indicate. That party must present the check to your bank to receive payment. Of course, if you live in Toledo, Ohio, and you send a $25 check to your cousin in Rapid City, South Dakota, she's not going to fly to Toledo to collect her money. So how does she receive payment?

The process has two parts: (1) Your check travels, either physically or electronically, through a series of financial institutions until it reaches your bank, and (2) credit for the amount of the check travels back through the system until it reaches your cousin's account.

Two of the institutions that form part of the sequence are the Federal Reserve banks closest to your and your cousin's banks. In this case, they are the Federal Reserve banks of Cleveland and Minneapolis, respectively. These large banks are part of the Federal Reserve System, which regulates U.S. banking and acts as a collecting agent for transactions between different cities. Your bank and your cousin's bank have accounts in these Federal Reserve banks.

In the first part of the process, then, your cousin deposits your check for a $25 credit to her account at the First National Bank of Rapid City. The Rapid City bank deposits your check for credit to its account at the Federal Reserve Bank of Minneapolis, which sends the check to the Federal Reserve Bank of Cleveland. The Cleveland Federal Reserve bank then sends the check to the First National Bank of Toledo, where you have your account.

The second part of the process begins when First National of Toledo deducts $25 from your account. The Toledo bank then authorizes the Federal Reserve Bank of Cleveland to deduct $25 from Toledo's Federal Reserve account. This amount is shifted from the Reserve bank in Cleveland to the one in Minneapolis. The Reserve bank in Minneapolis credits the account of the First National Bank of Rapid City for $25, and the Rapid City bank then credits your cousin's account for the same amount.

FIGURE 1–11
Process explanations (pp. 58–59).

Selecting Visuals

When you select visuals for a project or report, consider your audience first. A technical device or process may need clarification for a general audience but not for a technically sophisticated one. Visuals can help you explain things when such explanation is needed. Yet, even for a knowledgeable audience, visuals can increase variety and interest and dramatize differences. Your choice more often lies between types of illustrative materials rather than between using them or not.

Tables. Tables offer effective ways to organize data comparatively, so that readers can grasp and retain important points. In the table shown in Figure 1–12, for example, differences in the consumption levels of various liquids are shown in the

The cycle is completed when you receive notification, typically in the form of a cancelled check, that your account at the Toledo bank has been reduced by $25.

USING AUTOMATIC TELLER MACHINES

Most banks today provide automatic teller machines (ATMs) to serve customers at convenient times and places. Without incurring additional personnel costs, banks can handle deposits and withdrawals 24 hours a day, every day, through machines located in their own lobbies as well as convenience stores, shopping malls, resorts, and other areas accessible to the public.

For customers, one of the most popular roles of the ATM is to serve as a handy dispenser of cash. Making a withdrawal from an ATM is a process very similar to using a computer because the ATM is, in fact, a user-friendly computer. The cash withdrawal transaction involves four stages that are common to all computer systems. These are (a) input, (b) storage, (c) processing, and (d) output.

The *input* stage begins with the insertion of the customer's bank card into the machine. The bank card is coded with the customer's name and account number. This information must be supplemented with the customer's Personal Identification Number (PIN) to help identify him to the machine.

The identification data entered during ithe input stage are now *stored* in the memory of the ATM system. A second storage area of the system contains a file with the customer's actual PIN, his account balance, and other information.

The *processing* stage begins when the computer compares the identification data just provided with the information in the secondary storage area. If the PINs match, the customer is allowed to make another input by entering the amount he wants to withdraw.

In the case of a successful transaction, the *output* stage delivers the desired cash through a mechanical device to the customer. Several other outputs are possible, however. If the funds requested exceed the account balance, an "insufficient funds" message is provided, either on the screen or as a voice transmission.

In some cases, after repeated inputs of an incorrect PIN by a customer, a different kind of output may occur. The machine may swallow the bank card and sound an alarm.

columns under each year. On the other hand, trends in consumption are apparent by comparing numbers across the rows of the table.

Figures. Figures presenting graphs and charts can also show comparisons and may be more effective than tables in dramatizing trends. In these cases, a visual expression of the data substitutes for a numerical one.

Line graphs are effective at showing trends over time or changing trends in response to causal factors. Comparisons can be further heightened by using several lines within one scaled pair of axes. Figure 1–13 is an example of a line graph showing some of the same data as the table in Figure 1–12.

Bar charts offer alternate means of displaying trends. Several bars shaded differently can be used in either a horizontal or vertical format. Both formats are shown in Figure 1–14. Bars can also be stacked to show the cumulative totals of several parts. In a stacked configuration, bars can be shaded to show the proportional contributions of parts to a whole (see Figure 1–15 for an example).

Table 1
U.S. Annual Consumption of Liquids
(gallons per capita)

	1957	1967	1977	1987	1997
Milk	42.7	43.4	38.2	35.1	35.3
Juices	8.5	9.2	10.8	12.6	13.7
Coffee	29.6	30.6	27.8	26.3	24.2
Tea	7.1	6.9	7.8	9.9	12.2
Beer	32.2	33.8	34.5	37.1	38.3
Wine	12.8	13.0	13.9	14.2	14.6
Liquors	7.8	7.5	6.9	6.4	5.8

FIGURE 1–12
A sample table showing consumption data over time. Trends are evident in the numbers tracked across each row.

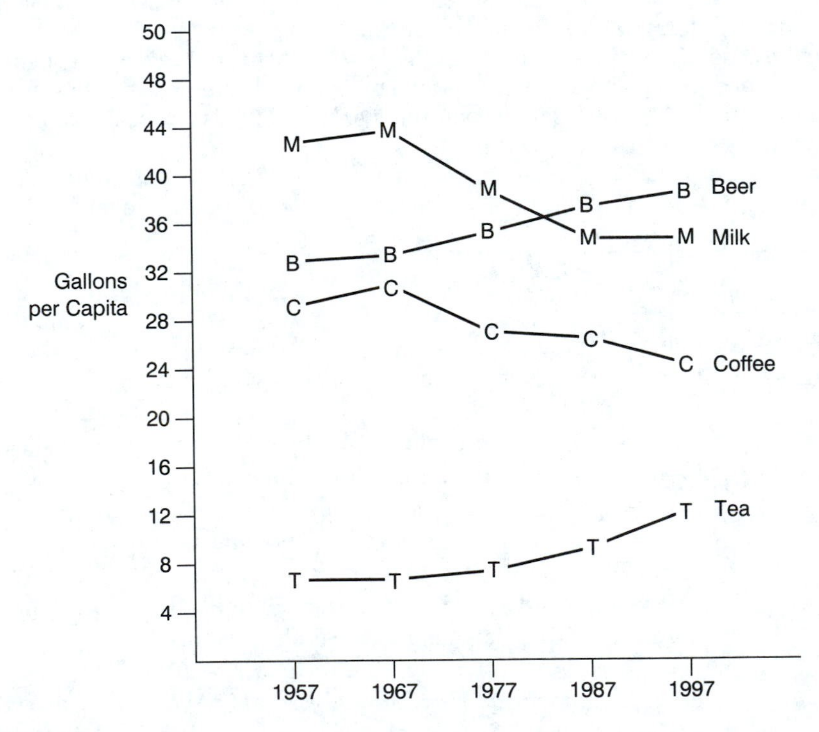

FIGURE 1–13
A typical line graph showing consumption data over time.

FIGURE 1–14
Two bar charts showing (top) horizontal and (bottom) vertical formats. Both show U.S. foreign trade.

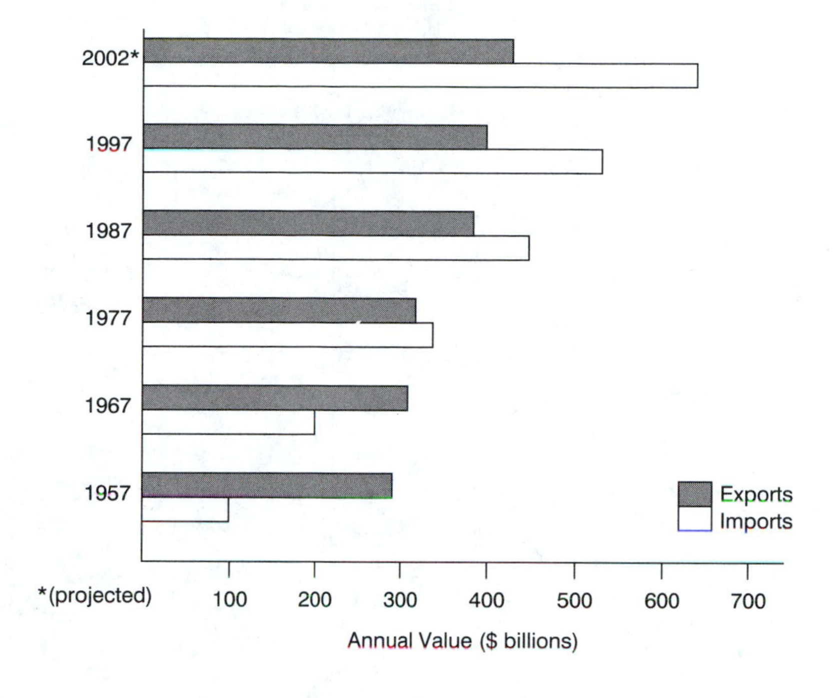

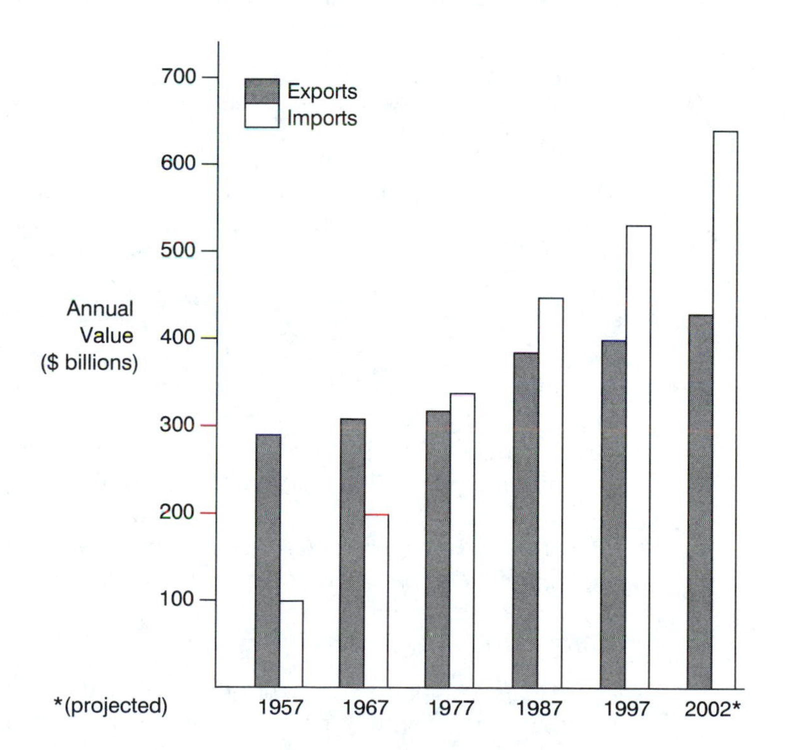

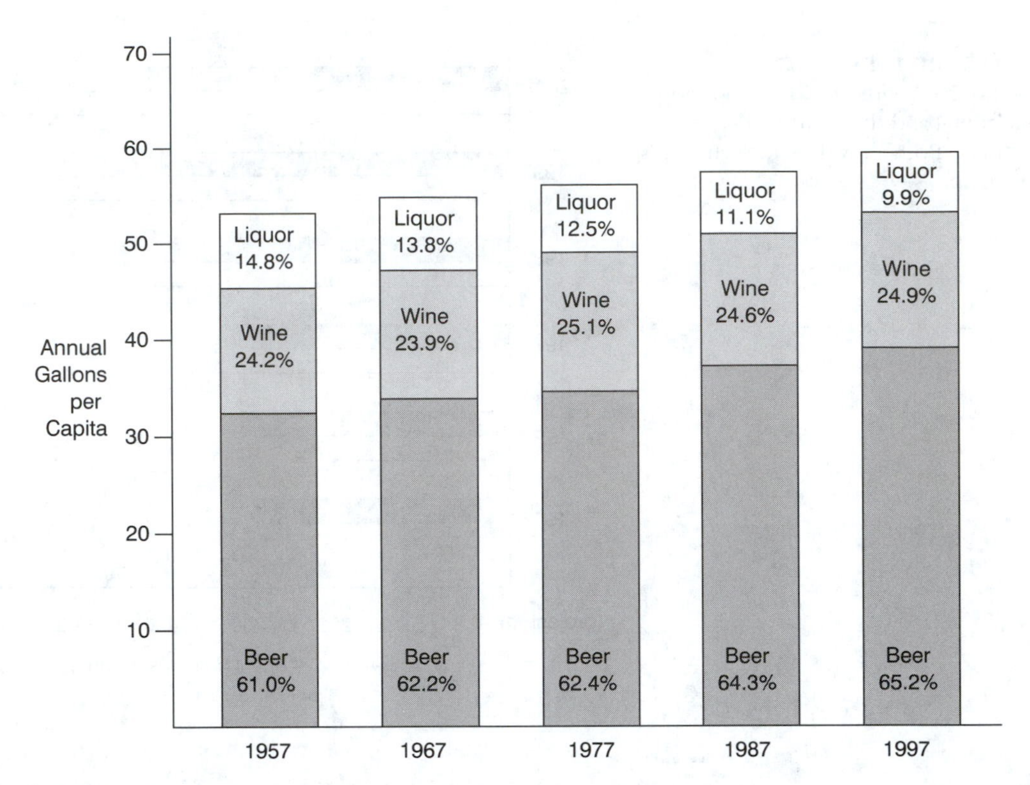

FIGURE 1–15
Stacked bars show proportional contributions to alcohol consumption data.

Pie charts can be used to show proportions and the relationship of parts to a whole. The largest part typically appears at the top, and the remaining parts are displayed in descending order of size, clockwise. Special effects can be added to dramatize key elements, as shown in Figure 1–16.

Flowcharts can show relationships and sequences within systems or processes. Typically, common geometric symbols available on templates (e.g., rectangles, parallelograms, diamonds, etc.) are labeled and connected with lines. The result is a schematic representation of reality that better illustrates the general condition than would a more specific illustration. See Figure 1–17 for an example of a flowchart applied to traffic violation cases.

Variants of the flowchart are the many kinds of project scheduling aids, whose purpose is to coordinate people and activities from several different areas over time. One of these is the *Gantt chart* (see Figure 1–18), which helps managers divide their time among a number of different projects with overlapping schedules.

Some Principles for Use of Visuals

Use visuals to support words, not to replace them. Practically, this means you should (1) refer to the figure or table with a specific comment about what it shows and (2) place the figure or table as soon after your reference as possible.

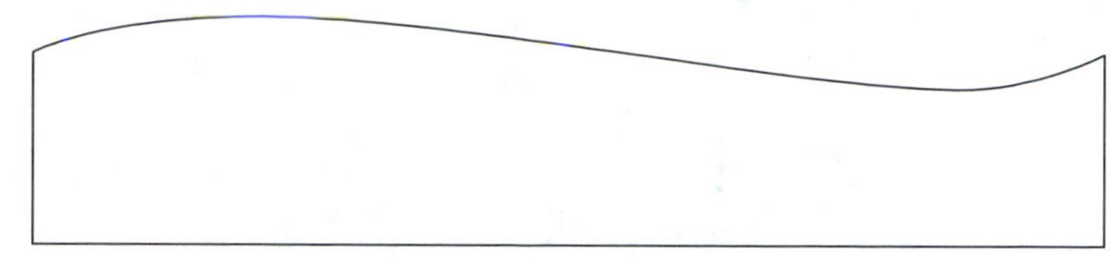

Figure 3. Digital transmission of analog signals using pulse-code modulation.

or

FIGURE 3 Digital transmission of analog signals using pulse-code modulation.

Table 2. Relative performance and general operation characteristics of major numeric coprocessors.

or

TABLE 2
Numeric Coprocessors:
Performance and Operation Characteristics

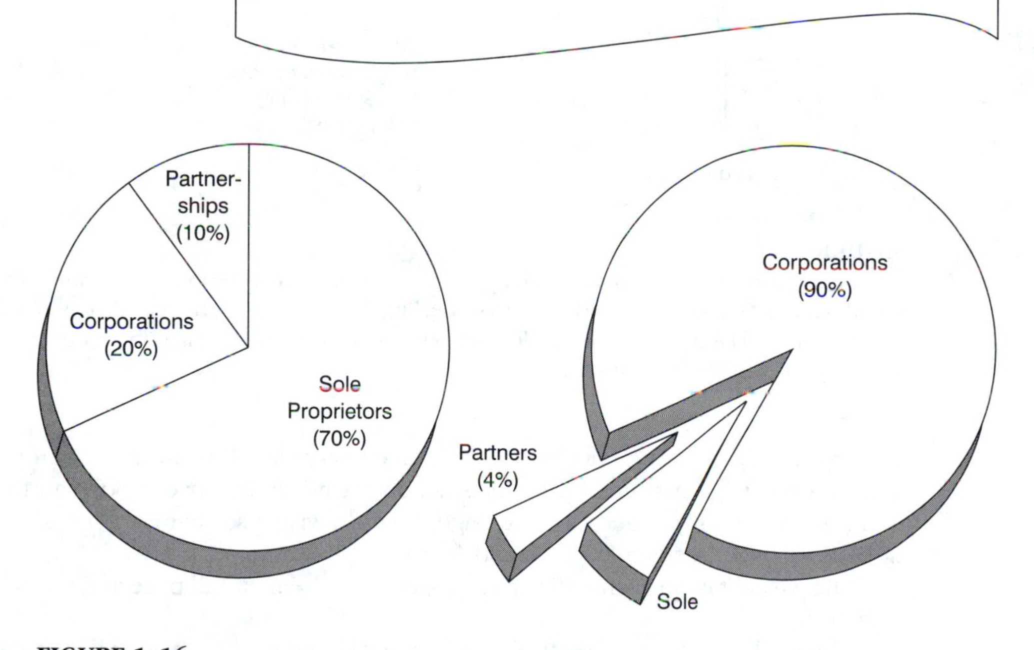

FIGURE 1–16
Two typical pie charts. The left pie shows the number of firms; the right pie shows revenues.

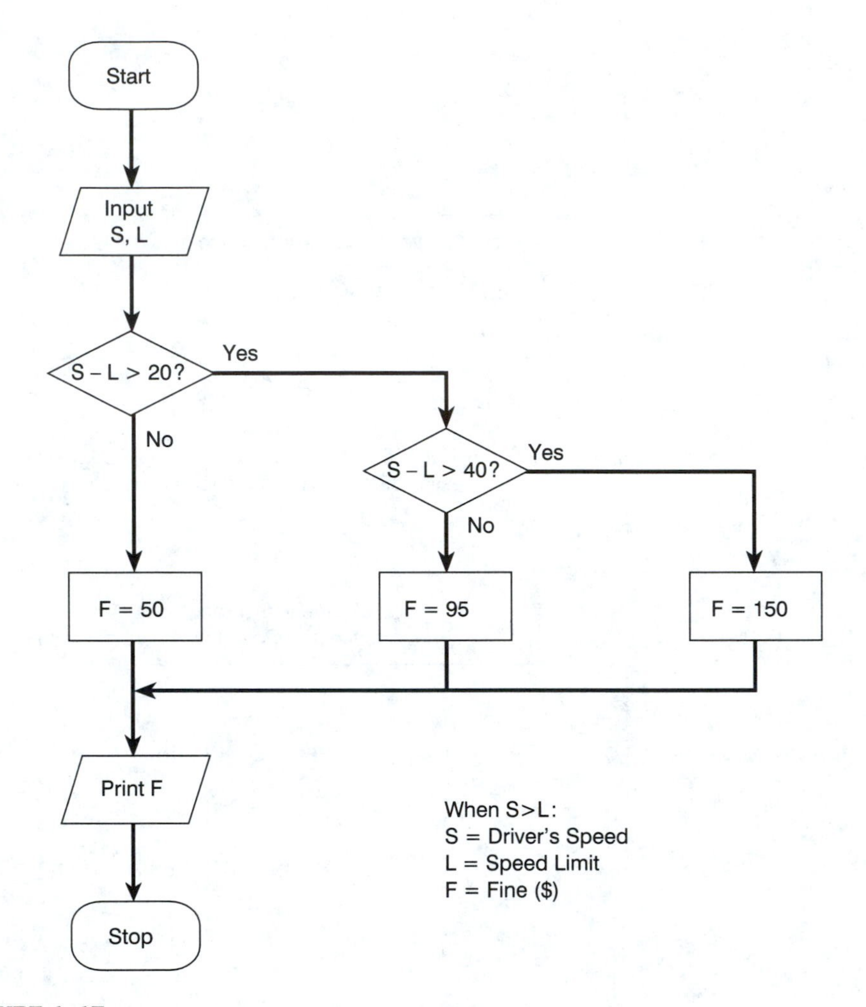

FIGURE 1–17
A flowchart showing the calculations made by state troopers to determine the amount of a speeding fine. Geometric symbols used for steps in the process include the "flattened" oval for the start and stop stages, the parallelogram for actions taken, the diamond for decisions, and the rectangle for calculations.

Make visuals large enough to show necessary detail. Generally, center them between the page margins on either side, and leave white space above and below them. If the visual is wider than the page, *turn* the visual so that its top is at the left side of the page (see Figures 1–14 and 1–19).

If a visual takes up an entire page, place it on the next full page after your reference to it.

Label all visuals, number them consecutively, and give each an explanatory caption. Figures may be labeled at the top or bottom; tables should always be labeled at the top. Use the form shown in the models at the top of page 63:

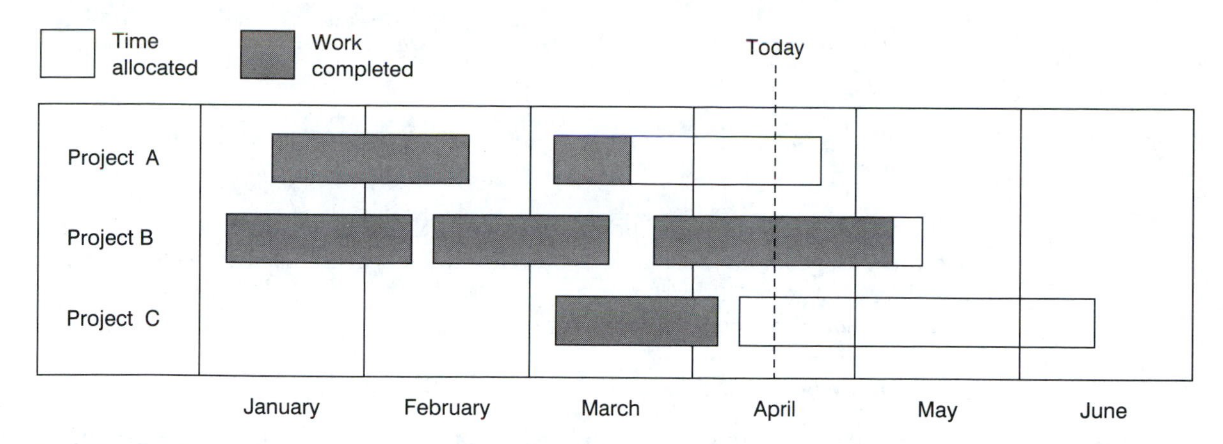

FIGURE 1–18
A Gantt chart showing the status of three projects for a manager who is ahead of schedule on Project B but lagging on A and C.

If you are not using a word processing or spreadsheet package, draft figures neatly, using straight edges, templates, compasses, and other aids as necessary. In tables, organize data into columns (vertically) or rows (across page). Label each column or row in terms of the items it contains and the units in which the items are measured:

Core Diameter (ml)

Don't separate columns or rows with lines, but space the items themselves to indicate a column or row. Align figures consistently, such as on the decimal point or on the right-most digit:

FIGURE 1–19
Placement of visuals on a page.

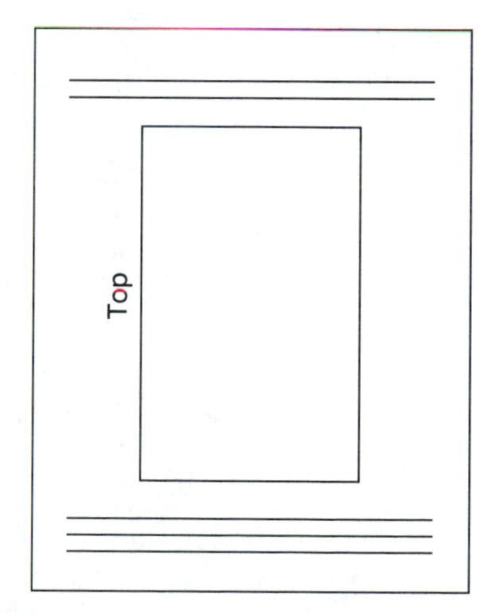

Temperature (°F)	*R & D Expenditures ($000)*
72.815	273
143.6	1,481
5.32	64

In most graphs, put the independent variable (the cause) on the horizontal axis and the dependent variable (the effect) on the vertical axis. This will be consistent with reader expectations.

(*Note:* See also Unit 4.6, Visuals for Oral Presentations.)

Unit 1.8 Exercise

1. Prepare a line graph from the following data: In 1988, 3.1% of college freshmen said they were planning to major in a computer field; in 1989, the proportion was 3.6%; from 1990 to 1996, the proportions planning to major in computers were 4.9%, 6.5%, 8.4%, 8.3%, 5.8%, 4.4%, and 3.5%, respectively. From 1988 to 1996, the percentages of freshmen intending to major in engineering were 10.3, 10.6, 11.8, 12.0, 12.6, 11.7, 11.0, 10.7, and 10.9, respectively. The percentages of freshmen declaring accounting as their intended major were 6.7, 6.2, 6.2, 5.8, 6.2, 6.3, 6.4, 6.5, and 6.2, respectively, from 1988 to 1996.
2. Use the data in Question 1 to construct a table.
3. According to the U.S. Bureau of Labor Statistics (*Occupational Outlook Quarterly,* Spring 1992), the engineering job outlook from 1990 to 2005 is as follows:

Type	***1990 Employment***	***Percentage Increase by 2005***
Aerospace	73,000	20
Chemical	48,000	12
Civil	198,000	30
Electrical/electronics	426,000	34
Industrial	135,000	19
Mechanical	233,000	24
Metallurgical, ceramic materials	18,000	21
Mining	4,200	4
Nuclear	18,000	0
Operations research	57,000	73
Petroleum	17,000	1
Systems analysts	463,000	79

   a. Use these data to prepare a bar chart of 1990 engineering employment, arranging the bars in order by the number of jobs. Select either a vertical or horizontal format, and be prepared to justify your choice.

   b. Prepare a bar chart with stacked bars, showing 1990 employment and the projected increases in the number of engineering jobs by 2005.

4. Tax revenues collected from various sources by federal, state, and local governments in the United States in the mid-1990s amounted to the following percentages of the total: individual income taxes 32%; corporate income taxes 13%; payroll taxes (social security and indirect business taxes) 27%; sales and excise taxes 15%; property taxes 10%; and other taxes 3%. Prepare a pie chart showing these data. Justify your ordering of the segments in the pie.
5. Prepare a Gantt chart including the major projects in all your classes for the remainder of the semester. Put the chart on the wall above your desk. Use it!

SECTION 2

Reports

Common types of reports are described and illustrated in this section. They are arranged in order from the shorter and simpler to the longer and more complex. Length and complexity, however, are not strictly proportional in a report. Complexity is mostly a function of two elements that may also be present in relatively short reports. The true complicating factors are (1) the structure and (2) the level of analysis or evaluation.

Reports are written for specific, practical purposes. They are read by busy people in the course of their work. They deal with ongoing problems and issues. For all these reasons, structure becomes an important consideration.

The way a report is organized, sequenced, and physically laid out—its *format,* in other words—can help achieve its purposes. Good organization and layout help the writer present the subject and emphasize the important points. A logical and efficient structure also helps the reader get through the report and focus on the main points.

The recurring, day-to-day situations that create a need for reports also create expectations about the format of reports. Field reports and progress reports, for example, are expected at regular intervals and in a predefined form so that supervisors can monitor activities and identify problems. Lab reports and user's manuals carry certain format expectations because of their subject, purpose, and audience.

The format generally becomes more elaborate if the report is destined for an outside audience. While many problems within an organization can be addressed through informal reports, or focused reports with a relatively simple format, outside reports must respond to a wider range of audiences and needs. They must provide background, cite sources and other evidence, and deal with liability issues. The result is a more complex structure, with additional front-end material, subordinated sections, and back-end attachments. Unit 2.7, Proposals, illustrates this difference through its formal and informal versions.

A report's complexity may also increase because of its level of discussion. The reports at the front of this section are designed primarily to inform readers. Later reports are more complex because they center on analysis, evaluation, and persuasion.

The development methods recommended in these units are logical ways to approach problems or issues through particular kinds of reports. The formats recommended for these reports offer logical and commonly used ways of organizing such tasks. As you gain experience, you may develop other, equally effective, approaches. Furthermore, instructors and supervisors may require your reports in different formats than you learn here.

Such differences are likely to be slight, however. Your understanding of the basic purposes and forms of reports, as well as your experience in writing these reports, will help you adjust to new requirements.

UNIT 2.1 INFORMAL REPORTS

In the business world, the informal report is one of the most common types of writing activities. Ideas, suggestions, clarifications, updates, problems, and other issues are shared with co-workers. The message is committed to writing so that the subject is stated clearly, sources are documented, and information is not lost.

Although they are informal, reports of this kind should be treated as important assignments. They are a record of your contributions to the organization. They deal

POSSIBLE ASSIGNMENTS

Briefly report on one of the following:

- ❑ The findings from a laboratory experiment
- ❑ The status of a course project
- ❑ Your observations on a class presentation
- ❑ Your recommendations for improving a class or lab procedure
- ❑ Information from research into a specific hardware or software product
- ❑ An accident you observed or experienced on campus
- ❑ An informational interview on career opportunities at a firm
- ❑ Information on products or technology applications gained from an article or manual

Approximate length: several paragraphs to several pages.

with problems that could grow if ignored. Exercises in informal reporting during college will help you produce effective reports later in your career.

Explanation of an Informal Report

What makes a report "informal" rather than "formal"? Informal reports usually convey routine, schedule-related information, such as the details of a project. They are often a quick response to a request for information, as when a supervisor asks an employee to clarify a customer complaint or when one department requests data from another. Informal reports do not contain the front-end or back-end components found in formal reports; there's no cover page, abstract, table of contents, bibliography, or appendix. Informal reports are the day-to-day practical vehicles for keeping an organization's members in touch with one another.

Because of the emphasis on timely information, informal reports can take a variety of forms. Depending on the purpose of your message and the circumstances, you may report through a letter, memo, preprinted report form, or e-mail. You might also report by attaching a copy of the requested information (a diagram, a table, an article) to an explanatory memo or letter.

Because the circumstances of informal reports are so varied, there is also no set way to write them. A memo report, for instance, could be a single sentence. A report attached to a memo or presented in a letter could be pages long, with subheadings and illustrations. Both are "right" if they fit the need.

The need to be met is to present information so it is meaningful to a reader. Be clear about what you need to say, and then say it with your reader in mind. For example, don't use technical terms a nontechnical reader won't understand. Don't spend time introducing a topic if the reader is already familiar with it. Get right to the point. Write with a purpose. Anticipate questions your reader might have.

In planning your informal report, refer to the developmental strategies summarized in the checklist on pages 29–31. While the subject of the report may seem self-evident, that is, derived directly from a specific situation or assignment, a review of your purpose and the likely audience may be helpful. It would also be a good idea to check the accuracy and completeness of the data you use and the soundness of your inferences.

The next section shows two types of informal reports written under different circumstances.

Examples of Informal Reports

A Report in Memo Form. Linda Wilkins, an operations manager at the Countryside Electronics Company, was reading a summary of the firm's monthly accounts when she noticed that charges for customer equipment damaged during service calls had increased. She called the supervisor of the service department to find out why. After learning that the majority of cases involved accidental short-circuiting of equipment, Wilkins sent a memo to the company's maintenance chief and asked for his advice on fixing the problem. The memo in Figure 2–1 is the maintenance chief's report to Wilkins (with a copy to his supervisor, Sharad Jani).

INTEROFFICE MEMO

TO: Linda Wilkins FROM: Don Gioretta

c: Sharad Jani DATE: November 11, 1996

SUBJECT: Short Circuit Damage

An increasing number of manufacturers now use static-sensitive equipment in their plants. Accidental short-circuiting during our troubleshooting procedures can frequently be traced to the static sensitivity of these devices. By purchasing antistatic kits and supplying our field personnel with them, we can reduce the level of short-circuit damage.

I recommend the RCA Antistatic Kit, which retails for about $54. It is the most effective and readily available kit on the market for its price. The kit is compact, easy to work with, and comes with a good set of instructions.

FIGURE 2–1
Sample memo report.

An Informal Report on a Problem. John Rosenthal's supervisor asked for information on one of the issues arising from a site selection project John was working on. John sent the report in Figure 2–2 attached to a cover memo so the report could be easily copied and added to the file on the project.

Unit 2.1 Case

You are the team leader of a group of five last-term computer information systems students working on a project for the Wheaton (IN) Historical Society. Your task is to computerize the society's archives and related records for the benefit of researchers, society members, and the general public.

The project had started well, but as the end of the semester and graduation approach, you find team members growing more lax in carrying out their assignments. Three weeks before graduation, team member Donald Boone tells you he's sorry but he will not be able to complete his programming task because Sandra Stiller, his partner, has not provided the necessary lists of documents. You call Sandra, who says she didn't know she had to compile any lists and resents being accused of not doing her job.

Without the Boone-Stiller component, the project will be incomplete, but neither you nor the other team members have time to make up the missing piece along with their own tasks and the requirements of other courses.

As the time approaches for your final status report on the project, you are torn between pointing the finger at your teammates and accepting responsibility yourself. "I *am* the team leader," you remind yourself, "and the society was depending on us." Since you will be working in Wheaton after graduation, you wonder if you should promise to help complete the project then. "I could do it, but would that be a cop-out from my role as team leader?" you ask yourself.

Write a status report to Wilma Cather, president of the Wheaton Historical Society, with a copy to your instructor, Professor James Landis. Let them understand what has happened, and tell them what they can expect from the project by term's end and beyond.

UNIT 2.2 FIELD REPORTS

Many work situations require communication that is highly concentrated—brief, to the point, condensed to its essentials. People who log in telephone messages and programmers who provide comments alongside their programs must write condensed messages regularly. One of the most common applications of this technique is in the field report filled out on a prepared form by a service technician, troubleshooter, or sales representative.

BRANCH STORE SITE SELECTION—PROJECT NO. 3560

Site C: Local Traffic Impact

Site C is currently served by one major artery (Randall Ave.) at the north end of the available acreage. There are only local-access roads on the other three sides. To make the store more accessible to shoppers, we would need to extend Lehigh Ave. for through traffic east and west. Lehigh is now interrupted at McDougal Rd. by a sanitary canal, and a bridge would be required to extend the road. While city officials have assured us they support the extension, area residents have expressed strong reservations. They fear increased noise, congestion, and pollution.

The issue may come to a head March 14, when Randy Soberg and I have been invited to make a presentation at the Elmwood Neighborhood Association meeting. I will keep you advised of developments.

FIGURE 2–2
Informal report.

In all these cases, space and time are at a premium. The ability to get to the heart of the matter and communicate only the essentials is needed.

Writing Condensed Field Reports

Frank Cilento works for a power company. One stormy night, a power outage is reported in the 5700 block of LaGrange Street, and Frank is sent to investigate.

When he arrives at the scene, he sees a group of people milling around a truck that has crashed into a utility pole. The driver appears to be in shock, and one bystander tells Frank that an ambulance has been summoned.

The utility pole is listing, with a severed voltage line dangling from it. The line is flapping wildly in the street, sending showers of sparks and loud electrostatic cracks through the air.

Frank rushes to contact his supervisor. Unfortunately, the phones lines in the area are out too, and Frank is driving a backup vehicle without a mobile phone. He does have his laptop computer in the car. The laptop is equipped with a wireless modem that allows transmission through cellular means. He decides to send a fax from his car. Realizing that his supervisor will be flooded with requests for emergency service tonight, he decides to make his message as brief and to the point as possible. This will be the best way to secure a response.

If you were Frank, what would you say in your message? What essentials of the situation does the supervisor need to know to respond?

Here's the message Frank sent:

> Turn power off 5700 block LaGrange. Live wire down. Cilento.

POSSIBLE ASSIGNMENTS

Condense one of the following to a preset length:

- ❑ A description of a faulty circuit, program, or system and the actions taken to correct the problem
- ❑ Program comments to help other programmers understand the logic of a program section
- ❑ Results of a telephone inquiry about products or services
- ❑ A field service problem along with corrective actions taken
- ❑ A request for help or information from a co-worker

Approximate length: A few lines to a few paragraphs.

Note: After you study the explanation and examples, try your hand at writing a condensed report in the form at the end of the unit (see Figure 2–6).

Frank managed to convey the necessary parts of his message in a very limited number of words. Your objective in this unit is also to reduce information to its essential elements so that the message gets across in the quickest, most direct way.

One technique for condensing information is to use the top-down method of newswriting. The first paragraph of a news story is often a single sentence that contains the who, what, when, where, and why of the story. An example might be: "President Smith [who] vetoed a tax bill [what] today [when] at the White House [where] because it was 'too extravagant for a nation seeking to put its fiscal house in order' [why]."

Details are added in following paragraphs, in order of diminishing importance. Editors later cut stories to size from the bottom up, on the assumption that no crucial information will be eliminated until the first paragraph is reached (see Figure 2–3).

Besides trying to get the "5 Ws" into a single sentence and adding details only as space allows, you can also use abbreviations, phrases in place of sentences, and other informal wording to save space. You can omit articles (*the, a, an*), conjunctions (*but, and, or,* etc.), prepositions (*of, on, in, to,* etc.), and forms of the verb *to be* (*is, are, was,* etc.).

Remember Frank's message? In full, formal English it would read something like this:

> Turn *the* power off *for the sector that includes the* 5700 block *of* LaGrange *Street.* A live wire *has been severed and is dangling* down *to the street. Frank* Cilento.

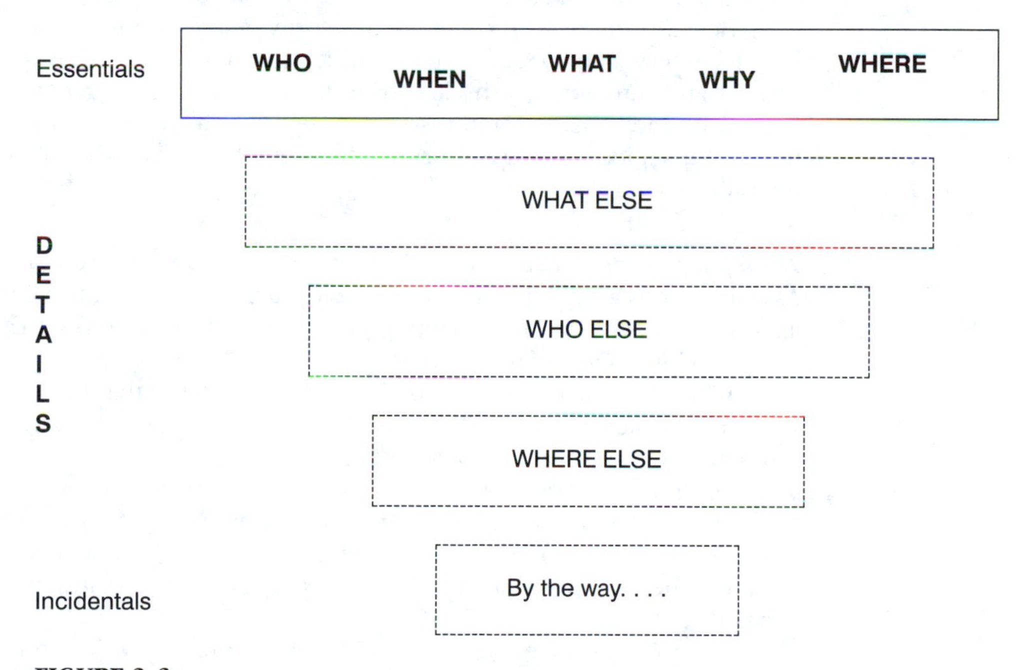

FIGURE 2–3
Structure of a field report: The top-down, or order-of-importance, method used in news stories gives all the essential information in the first paragraph.

A final technique that might be effective is to pretend you are sending a telegram at $10 a word. You want to save money, but if you leave out any essential information, the receiver of the telegram will charge you $100 for each missing detail. In other words, although being *concise* is important, being *complete* is still the most critical factor in writing a condensed message.

Examples of Field Reports

The Access Control Problem. Mammoth Data Systems had designed a large centralized computer system for Creditworthies Inc., the national consumer credit investigation agency. A mainframe computer in the credit agency's headquarters in Omaha is linked to remote terminals in client banks and loan agencies through a nationwide communication system. Information on consumer credit activities is entered into the mainframe or accessed from it via the remote terminals and communication lines.

With thousands of client terminals in use, and possible access to the central data base from unauthorized sites, security of the system and confidentiality of its data have posed an increasing challenge. In several instances, lists of consumers using or applying for credit appear to have been generated from the Creditworthies data base and sold to clients. Vice President Ripley of Mammoth Data asks Ken Janacek, a systems analyst, to study the security issues and provide short-term and long-term recommendations for improving the system's security.

Because the need for upgraded security is a pressing one, Ken visits several sites to observe the conditions under which the system is used. Based on his observations and interviews with clients, he formulates a set of recommendations for immediate measures to tighten security. He writes a summary of his recommendations on a standard company form and sends a fax transmission of it to Vice President Ripley (see Figure 2–4).

The Field Service Case. Myra Tillman, field service representative for Hornsby Associates, was assigned to investigate a customer complaint. Clive Bates, office manager at the Washburn Manufacturing Company, had called to say that his computer was scrambling some of his customer accounts.

When Myra arrived at Washburn, Mr. Bates told her that the computer, an IBM PS/2, had worked well for over 2 years. It was used for customer accounts and inventory control. A week ago, problems had developed: Inventory results and accounts were intermittently being reported in completely scrambled form.

Myra ran a diagnostic test on the computer. The test enabled her to locate a faulty RAM chip. After replacing the chip, she reran the diagnostic program and obtained the proper results. Finally, the accounts were run again, and all came out correctly.

The job was done, but Myra still needed to fill out a field service report and have Mr. Bates sign it before she left. A part of the form as Myra filled it out is shown in Figure 2–5.

Now try writing a condensed report, using the form supplied in Figure 2–6.

MAMMOTH DATA SYSTEMS

ACTION ALERT: **For Immediate Consideration**

Route to: ________

TO: Harold Ripley

RE: Creditworthies Syst. Security

Access to our data base is now too open. Suggest installing call-back system in Omaha to make access possible only from client terminals. Longer term, may be advisable to install security software to scan passwords and IDs, limit user ability to get certain kinds of reports.

8-6-96	Ken Janacek
(Date)	(Sender)

FIGURE 2–4
A condensed report in the computer system security case.

UNIT 2.3 PROGRESS REPORTS

Because progress reports are important tools of management control, they are often a requirement in major projects as well as a means of tracking day-to-day activities. A key management concern is to identify problems early so that adjustments can be made and greater problems averted.

Customer Comments and Repair Details

Customer complained of intermittently scrambled results in accounting program in IBM PS/2 Model 95. Test confirmed complaint. Found RAM chip with bit stuck high. Replaced U50 (16K RAM). Unit now meets specs.

Supervisor Approval	Safety Check	Customer Signature	Date
	3/21 – MT.	X C.Bates	3/21/96

FIGURE 2–5
Myra's field service report.

Description of Problem

____

____

____

Action Taken to Correct Problem

____

____

____

FIGURE 2–6
Sample form for condensed messages. (*Note:* Write only on the lines provided.)

For the writer, the advantages parallel those of management. Planning the report gives you a chance to review your activities and to identify necessary changes. The report itself is an opportunity to present yourself positively, particularly as a problem solver. Your report provides at least the first step—problem identification. Your description of the circumstances and of the efforts you have made to deal with the problem may also contribute to an eventual solution.

POSSIBLE ASSIGNMENTS

Write a formal or informal progress report on one of the following:

- ❑ A series of related lab experiments
- ❑ Major design projects
- ❑ Projects to computerize business records or procedures
- ❑ Accounting or financial planning projects
- ❑ The development of a major writing project, such as a formal report, feasibility study, or user's manual
- ❑ Projects to implement new procedures or organizational changes
- ❑ Major equipment purchases
- ❑ Everyday activities over a 2- to 4-week reporting period

Approximate length: one page for internal reports; one to four pages for reports to clients.

How to Write a Progress Report

The specific form of a progress report depends on its type, its uses, and especially its audience.

The periodic progress report is a common device for keeping track of day-to-day activities. The reporting period may be a week, 2 weeks, or a month, and the report is addressed to a supervisor, who may then draw from it in reporting to the next level of supervision. Such periodic internal reports are usually submitted in a memo and follow a specified outline.

Progress reports are also required for important projects that extend over months or even years. The reporting period in such cases may be a month or a fiscal quarter (3 months). Such progress reports are more often presented formally, with a cover memo, a title page, discrete sections with subheadings, and even a table of contents and appendices. The report may be submitted to senior management or to a client firm for whom the project is being done.

The form of progress reports on major projects is often determined by the nature of the work being done. Progress on a construction project, for example, may be tracked by the completion of key structural parts such as the foundation, the external frame, the roof, and the like. Projects such as laying cable or building a road may be tracked by linear distances covered, that is, milestones reached. Design projects may be divided into phases such as research, schematics preparation, prototype development, and testing.

Apart from such considerations of type and use, a progress report can be described in general terms because it always involves the basic elements of work and time. Putting these elements together, the report must describe the following:

1. What has been done
2. What is being done
3. What will be done

When a particular method of reporting is not specified, the following would be effective in most cases:

The All-Purpose Progress Report

- **A.** Heading: identifies project, purpose, time period
- **B.** Introduction: previous work completed
- **C.** Present status
    - **1.** Present work
        - **a.** Work completed
        - **b.** Work started
    - **2.** Problems
- **D.** Work remaining
    - **1.** Work planned next
    - **2.** Assessment of progress
        - **a.** To date
        - **b.** To completion

For a formal progress report on an important project, a modified version of this approach is suggested. Because top management will read such a report, you should use an "executive summary": Simply move the last section—assessment of progress—to the front, directly after the heading.

A few explanatory comments on the all-purpose approach follow.

The *heading* can be given in the "subject" line of the memo form:

SUBJECT: Biweekly Progress Report 11/4/96–11/18/96

A "heading" can also be given on the title page of a more formal report:

Progress Report: August, 1996
LOCAL AREA NETWORK CONVERSION
for
Standalone Industries, Inc.
Contract No. 48312Y

The *introduction* should summarize the previous work done. If you're reporting monthly and will cover August in this report, briefly summarize work completed in July. A simple list may be sufficient.

The *present status* section describes activity during the current reporting period. What parts of the project were completed, and what work is in progress? Then point out problems in both completed and ongoing work, even if the problems have been resolved. Awareness of problems is important to management and can help you gain additional support for the project.

The *work remaining* section looks both forward and backward. After a brief list of work planned for the next period, the key element is your assessment of overall progress, primarily in terms of the schedule you are following. Is the project on schedule now, and will the entire project be completed by the final date? If you cannot say "yes" to both questions, think a bit about the tone of your explanation.

In most cases you will be trying to inspire confidence by the assurance that you expect to get back on schedule by a certain date. On the other hand, if the problems that caused the delay are still present, you may wish to seek help in removing or minimizing them. So point out their threat to the completion of the project. Your readers will undoubtedly prefer to hear about potential threats now rather than actual damage incurred later. In sum, your assessment should be positive but also realistic and forthright in looking squarely at potential delays.

Example of a Progress Report

Figure 2–7 shows a progress report on the development of a user's manual.

UNIT 2.4 FORMAL REPORTS

Many government agencies and public or private organizations and virtually all businesses that engage in contractual agreements produce formal reports of one type or another. A formal report is an extensive presentation of facts and ideas in a highly organized format. Its purpose is to inform or convince the reader through the logic

INTEROFFICE MEMORANDUM

To: Pat Drynan

From: James DeCarlo

Subject: June Progress Report
K-9820 User's Manual

Date: July 1, 1996

In May, we completed the following steps per the project plan authorized in February:

1. Interviewed R & D (Hodges, Klein, Metzger, and Alvarez) to gather technical information on the K-9820.
2. Interviewed Marketing staff (Roberts, Bender) on projected sales, dealer training program, and customer characteristics.
3. Drafted a preliminary outline and graphics plan for the manual.
4. Developed a working version of the table of contents.

Current Activity

This month, we made final assignments to the writers and editors (see Table 1) and to the art department, and completed drafts of sections 2.1, 2.2, and 3.1. These are now in editing.

Work in progress includes first drafts of sections 3.2, 3.3, 4.2, and Appendix A. Also, photos of the main K-9820 subassemblies were taken and are in processing. Of the projected 13 schematic diagrams, we have been able to obtain sketches of only 7, which are being redrawn in the art dept.

The remaining six schematics have been delayed pending "design improvements" in R & D. I am attempting to determine the extent of these changes and to estimate the length of the delay.

FIGURE 2–7
A progress report (pp. 81–82)

of its ideas and the quality of its presentation. A major purpose of the structured format is to give the information and ideas in the report the strongest possible support.

Learning to write a formal report will show you how to develop your material in depth. It will also give you a chance to construct a format by which these ideas are presented most effectively.

POSSIBLE ASSIGNMENTS

Prepare a formal report on a project or topic such as the following:

- ❑ A microprocessor-based control system designed for a specific application
- ❑ A computer-controlled stereo tuner employing a stepper motor

2

Work Remaining

In July, we expect to complete drafts of all sections started in June as well as sections 4.1 and 4.3. We expect to begin work on all of the remaining sections (1, 5.1, 5.3, 6.1, 6.2, and Appendices B to E). In the graphics area, I expect to complete approvals on the photos and to finish seven schematics and send these to R & D for approval.

Because of the delayed schematics, we are 2 weeks behind in our graphics program, while writing and editing are on schedule. If I receive the six delayed schematics within 2 weeks, I can get graphics back on schedule by the end of July via overtime for two CAD operators.

My concern is not so much the delay in these sketches as the possibility that "design improvements" may be extensive and may require rewriting of completed sections of the manual or even changes in the basic outline and its strategy.

If the design changes prove minor and are received in 2 weeks, I am confident we can complete a draft of the entire manual by August 30 as scheduled.

TABLE 1 K-9820 manual writing assignments.

	Sections	*Draft Due*	*Sections*	*Draft Due*
Perini	2.1, 2.2	6/15	5.3	8/15
Bryce	3.2, 3.3	7/15	6.1, 6.2	8/15
Leftwich	3.1	6/15	1, 5.1, 5.2	8/15
Brewer	4.2	7/15	4.1, 4.3	7/28
Fencik	Appendix A	7/15	Appendices B, C, D, E	8/15

FIGURE 2–7, *continued*

- ❑ An independent telecommunications system between the branches of a major corporation
- ❑ Installation of a satellite video conferencing network for a geographically dispersed organization
- ❑ An internal electronic mail system for a large corporation
- ❑ A local area network project
- ❑ The social issues (pollution, safety, privacy, etc.) associated with an emerging technology
- ❑ A business problem based on case information and involving research

Approximate length: 6 to 15 pages of text, plus front and back matter.

Explanation of a Formal Report

The components of a formal report have evolved from the need for a final document that is comprehensive and complete. A description of these components follows.

Title Page. The first page of the report should give the title, the report number (sometimes required in companies), the author's name, the company (or school, class, and instructor) for which the report was produced, and the date on which the report is submitted. Other information can be added as necessary.

Abstract or Summary. This section may serve as the entire report for some readers if it is an informative abstract—a tightly condensed version of the actual information contained in the body of the report. The informative abstract is also known as an executive summary. A second kind of summary is the descriptive abstract—a description of what the body of the report covers. A descriptive abstract is used in the Introduction section below.

Table of Contents. The table of contents should list all the major components of the report except the title page and abstract. If subheadings are used in the body of the report, the table of contents should present these as well. Corresponding page numbers are listed across from each component or subtopic.

List of Illustrations. If more than four illustrations are used in the body of the report, the number and title of each should be listed, along with the corresponding page number. Both figures and tables should be included as illustrations.

Introduction. As the first of the three main components of the report—the body and ending are the other two—this section *introduces* the reader to the body of the report (Figure 2–8). It tells the reader the following:

1. What the report is about (subject and scope)
2. Why the report was written (purpose)
3. How the body of the report is developed

To accomplish the first objective, announce the subject of the report and define or describe it if necessary. Then explain how broad or limited your coverage of the subject will be.

The scope of coverage will be linked to the second question—why you wrote the report. For example, if your company asked for a report on recent developments in a certain technology, you would report only on those that relate to the company's business. You would do so because your purpose is to investigate whether any developments in this field might be used to improve the company's operations.

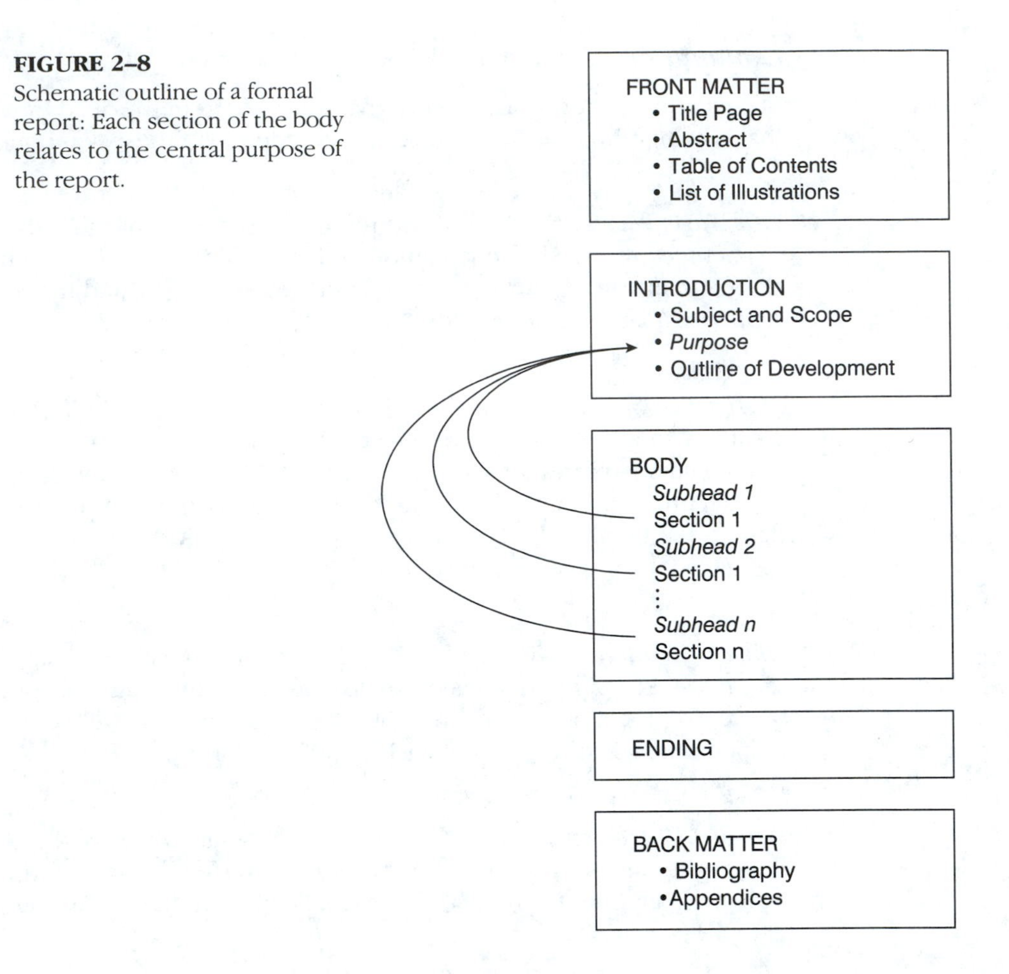

FIGURE 2–8
Schematic outline of a formal report: Each section of the body relates to the central purpose of the report.

For the third aspect of the introduction, list the major subtopics of the body of your report in the order of their presentation (i.e., develop a descriptive abstract). The topics can be presented within a sentence or two, or as a vertical list.

Body. The body is the largest and most important section of the report. All the other components are meant to support and lend credibility to this section.

How you develop and format the body depends on the type of report you are producing. If you are trying to convince the readers of something, present your argument so that you build point on point. In this case, you may want to avoid abrupt breaks and let the argument flow without many subheadings. If you're analyzing something, on the other hand, subheaded sections on each aspect of the subject would be appropriate and effective.

Most technical reports do establish a series of subheaded sections to guide and orient the reader. Each section presents information on the subtopic but also ties the subtopic to the purpose of the report. To establish these subtopics, you may need to provide definitions or background information in some cases.

The body may also include illustrations if you decide these are helpful. If you use an illustration, make sure you number it and refer to it by number in your text before it appears.

Ending. How you end your discussion also depends on the type of report you are preparing. The following are common endings:

1. Summary. If the purpose of your report is to inform, providing a summary or recap of the important points would be most helpful to the reader.

2. Conclusions. If your report analyzes something, you would logically reach conclusions about cause and effect. Restate your conclusions at the end to emphasize them.

3. Recommendations. If the point of your report is to evaluate something or to provide the company with information so it can make a decision, give your own recommendations here.

Bibliography. All sources of information that you consulted in preparing your report should be listed in alphabetical order here. (See Unit 5.4, Bibliographies and Citations.)

Appendices. For large diagrams and graphs, flowcharts, computer programs, and other data that are not easily incorporated into the body, sections labeled "Appendix A," "Appendix B," and so on, can be added at the end and referred to in the body.

A Logical Approach to Developing the Report

Before you begin a major project such as a formal report, review the developmental strategies and approaches suggested in the Introduction, and use the checklist on pages 29–31 to help you implement these strategies. When a formal report must be developed within severe time constraints—that is, under the typical conditions in school or on the job—a more specific application of the general strategy may be needed.

There's no telling how much time you'll be given to prepare a report, but the following approach will help you use your time most efficiently.

Week 1. Determine the *subject, scope,* and *purpose* of your report. In school, instructors may request that you submit a subject idea for approval to ensure you haven't selected a topic that is too broad or narrow, that would be too hard to research, or that isn't appropriate. On the job, your subject will usually arise from your work or will be assigned to you, but you'll need to clarify your purpose and decide on your scope. Ask yourself questions like these: "Why am I writing this report?" "Who is my audience?" "Am I writing to inform or to persuade?" "Do I need to provide background information on the subject?" Having a clear scope and a clear purpose may be the two most important factors in producing a good formal report.

(*Note:* When the project is an *assignment* by a supervisor or instructor, you still need to answer the first of these questions—"Why am I writing this report?" How might this question be sharpened to show why it requires some thought?)

Week 2(a). Decide *how you will gather the information* on your subject. Will you need to get information from other people, inside or outside the company? Will you be studying equipment or procedures in the lab? Will you be using published sources like books, magazines, data manuals, or electronic retrieval systems? Make sure you have a clear plan of action and an idea of how long it will take. For published sources, including those accessible through CD-ROM or on-line data bases, go to the library and see what information is available. In school, your instructor may want to see a list of the published sources you plan to use or a plan for how you intend to get the information.

Week 2(b). Gather your *information.* If preliminary work has already been done, you should be able to focus the research to a greater extent now. As you uncover information on your subject, you may find you need to change the scope of the report to reflect ideas that hadn't occurred to you at first, especially if the subject is new to you.

Take complete notes as you gather information through surveys, reading, lab activities, conversations, and other sources. Four main types of notes are most common:

1. Summaries of information
2. Restatements of someone else's words
3. Quotes you might want to include in your report
4. Your own comments on the information you find

Week 3(a). Make an *outline.* Start by reviewing the information you have gathered. Then try a modified brainstorming approach to help you generate additional ideas and cluster information already obtained. Work toward an outline of the main points and subtopics you will cover. If the outline you develop reveals gaps in your coverage, be prepared to conduct additional research or analysis. Stay flexible as the shape of your report begins to emerge, and be prepared to revise conclusions if new evidence is found.

Week 3(b). Write a *rough draft* of the body of your report. Use your outline to make sure you cover all the important points in a logical order. Use your notes to develop the details of the main points.

Week 4. *Revise* your draft of the body to make sure the information is complete and the language clear. Add graphics or visuals you planned for earlier, or those you decide on now. Then prepare the *other components* of the report. As you revise, judge your work from the viewpoint of the reader. Make sure the report leads the reader logically through the subject while providing every assistance in understanding what you are saying. Check for accuracy in labeling and referencing illustrations, sources, and items in the table of contents.

Example of a Formal Report

Peter Goldberg, a research assistant at Able Corporation, discovered a need for a type of device manufactured by his company, but altered to suit disabled people. Peter made the modifications and came up with a workable prototype. He was asked by his supervisor to prepare a report on his project. Peter's supervisor felt that Able's executive management should be informed about the capabilities and the manufacturing feasibility of the device. Peter's report is presented in Figure 2–9 in schematic form.

Title page

MICROPROCESSOR-BASED RADIO ANTENNA CONTROLLERS FOR THE DISABLED

By Peter Goldberg
Research Assistant
Research and Development Department

for Able Corporation

October 11, 1996

FIGURE 2–9
Formal report (shown in schematic form, pp. 87–91)

Here Peter condensed the information in his report, including his evaluations, to guide readers through the main stages of his presentation.

EXECUTIVE SUMMARY

A survey of disabled radio operators in the tri-state areas showed that a need and a market exist for a radio-antenna controller with special adaptations. The problem most often mentioned in the survey. . . .

Here Peter listed the report components—including the five subheaded topics in the body of the report—and their page numbers.

TABLE OF CONTENTS

Peter listed two tables (not shown here) that showed manufacturing costs, three schematics, and a cutaway drawing of his device.

LIST OF ILLUSTRATIONS

FIGURE 2–9, ***continued***

Introduction: **Here Peter decided to provide background in the first sentence of his introduction. In the second sentence, he combined an indication of his scope with a statement of the report's subject. The third sentence also deals with the report's scope, and the last gives Peter's purpose, or why he wrote the report.**

In the second paragraph, Peter explains how the body of the report is developed.

Many disabled people use amateur radio equipment for communication but often do not have the strength or coordination to aim the antenna with conventional antenna controllers. This report describes the advantages, design features, and manufacturing costs of a newly designed radio-antenna controller that is operated by a microprocessor. It also compares this controller to others manufactured by Able Corporation. This analysis is presented to determine whether the invention of this device offers a manufacturing opportunity for Able.

The report presents survey information on disabled radio operators; a discussion of the problems and needs of disabled operators; the design of the microprocessor-operated controller, including schematics and specifications; a comparison of the new design to controllers presently manufactured by Able Corporation; and manufacturing costs of the new device.

Body of Report: **Only the subheadings and beginning statements of each subtopic are reproduced here from the main section of Peter's report.**

A Survey of Disabled Operators

A conversation with a disabled friend illustrated the special problems and needs of disabled radio operators, and led to the idea of using the capabilities of a microprocessor to replace human motor skills. A survey of 87 disabled radio operators was conducted by phone to. . . .

Needs of Disabled Operators

The most frequently cited problem in the survey was the difficulty of applying the amount of pressure necessary to hold down the switches on conventional antenna controllers. . . .

The Design of the Controller

By utilizing a 6508 microprocessor, a 2716 EPROM, a 6522 Versatile Interface Adapter (VIA), a 7574 Analog-to-Digital Converter, a stepper motor, and an application program . . . . (see Figure 2). . . .

Similarities to Current Able Controllers

Adaptations needed to convert presently manufactured antenna controllers to microprocessor operation are minimal . . . .

Manufacturing Costs of the Controller

The costs of components used in the microprocessor-operated controller are shown in Table 1 . . . .

CONCLUSIONS

Based on the informal surveys conducted during the design of the radio-antenna controller presented here, marketability of the design appears. . . .

Peter decided to end his report with his own conclusions about the practicability of manufacturing his device. He listed the positive aspects of his design—marketability, low manufacturing costs, and compatibility with current manufacturing operations—and concluded that an opportunity did indeed exist.

FIGURE 2–9, *continued*

Peter listed alphabetically all the sources he had cited, as well as those consulted in researching the report.

BIBLIOGRAPHY

Gunther, Alex. The Special Needs of Special People. New York: Putnam Books, 1993.
Hargrave, Shelley. Amateur Radio Guide: A Sourcebook. LaJolla, CA: Valley Publishing, 1995.

Peter used two appendices: Appendix A for a list of his survey questions and Appendix B for his application program. He had referred to both in the body of the report.

APPENDIX A

Survey of Disabled Radio Operators in Tri-State Area

1. How long have you been operating . . . .

UNIT 2.5 FORMAL LAB REPORTS

To write a good formal lab report, you have to grasp a technical problem as a whole and then organize your perceptions into an effective presentation. Instead of just following directions in a lab manual and filling in tables, you must step back and consider the purpose of your work and the implications of your results.

By using a specified format and a carefully structured approach, you are also learning about objectivity, precision, completeness, and other essentials of effective technical reporting. Your ability to give a precise, complete picture of your work to others will be a valuable asset in your career.

POSSIBLE ASSIGNMENTS

In a specified format, report on one of the following:

- ❑ One of the experiments or projects scheduled for a lab associated with a technical course

- ❑ A series of related experiments or projects

  Approximate length: three to eight pages, longer for complex projects.

Explanation of a Formal Lab Report

In the laboratory in industry, problems arrive unstructured and solutions are open ended. In the academic laboratory, lab problems are designed and their results are foreseen by the faculty. Because of this difference, results of academic laboratory work are often reported in a shorthand form in abridged formats. Students may submit only calculations and values obtained. Circuits may be visually inspected by an instructor, with no written materials required at all. Within heavy academic schedules, such shorthand forms of reporting are often a necessity.

In a broader sense, however, the academic and industrial laboratories have important similarities (see Figure 2–10). Both settings emphasize the scientific (experimental) method of testing a hypothesis and developing a systematic, logical approach to a problem. This method requires objectivity and precision so that problems can be explored, results tested, and conclusions drawn that are valid and generally applicable. In both cases, therefore, laboratory work is done with the expectation that it will be reported to others or at least checked by them.

The formal lab report is merely an extension of these common elements. In addition to a systematic approach, precise recording of results, and a spirit of objectivity, the formal lab report requires an introduction to the purpose of the experiment and some discussion of the results. These additional features are consistent with the need to inform others of the methods used and the findings determined.

Preparing the Report

In college, your task of writing a formal lab report is made easier by the fact that a highly structured procedure with specified equipment is usually laid out for you. By following the instructions in the lab manual and observing standard laboratory practice, you should be able to derive all the data and other information you will need for your report.

When you have completed the experiment and collected the necessary data, you can conduct an analysis and draw conclusions from the results. This part of the process may require you to perform additional calculations or develop graphs based on the data. You may also wish to evaluate the experiment and suggest practical modifications of the experimental procedure itself. You can then write up the analysis and conclusions as the last section of your report.

In writing the report, you should follow the format specified by your instructor. The format shown in this unit is a generally suitable approach for academic labs, but it may be modified for special conditions.

When you write the report, adjust your perspective from the lab manual you have been using to the report you are developing. Remember to use past tense language ("I adjusted the voltage to . . . ") because you are reporting what you *did.* Do

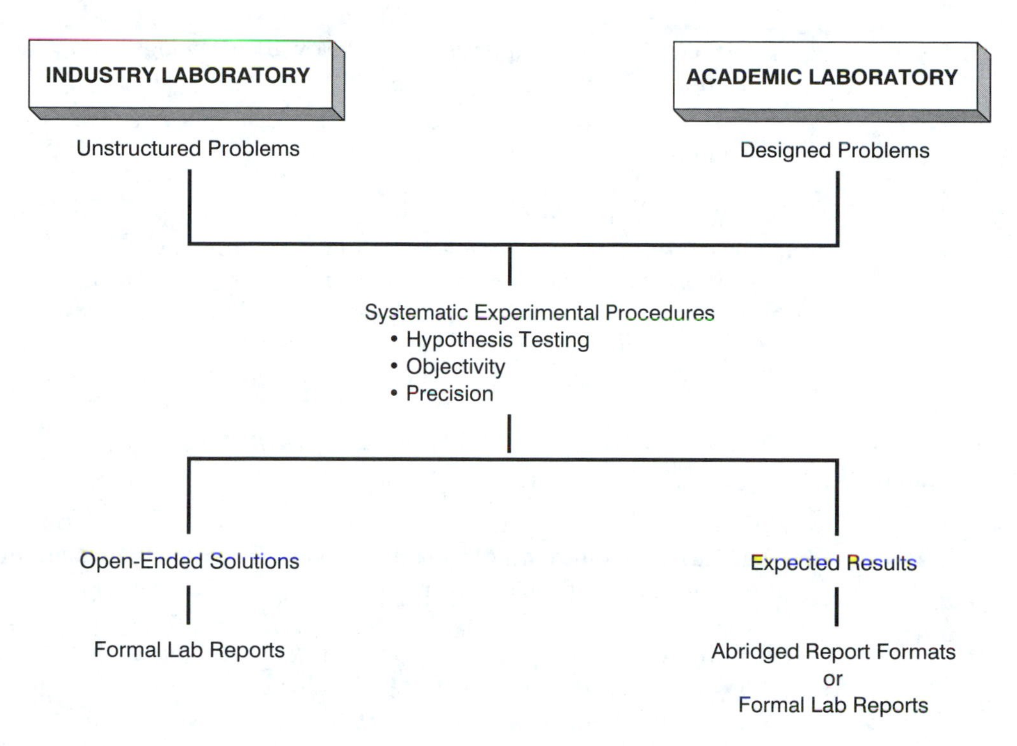

FIGURE 2–10
Both industry and academic laboratory settings require systematic procedures that lend themselves to review by others.

not use the language of the manual, which is written as instructions ("Adjust the voltage to . . . ") and may be in future tense ("You will find that . . . "). Use precise description that is supported by figures, tables, and graphs.

Because this is a formal report, you should take extra care with its appearance and general neatness. For best results, write a first draft of the text, and make notes on where you intend to include figures and tables. If you can use a word processor with a graphics package, prepare your illustrations next and import them into the text. Otherwise, sketch the illustrations freehand to get an idea of their size. Prepare explanatory captions for the figures and number them consecutively; then repeat this procedure for the tables, which should be numbered separately.

After you revise your first draft of the text to eliminate unclear writing and grammar and spelling errors, you are ready to produce the final version of the report. For best results, use the following guidelines:

- ☐ Type the report or key it into a word processor; leave room for the illustrations.
- ☐ Draft all illustrations neatly so they are large enough to show necessary detail, and center them across the page of your manuscript.
- ☐ Place schematics and tables in the text after they are mentioned, as close as practically possible to the references.

- ☐ Label circuits and graphs as *figures* below the drawings and include explanatory captions.
- ☐ Place data in *tables;* label these above the columns, and give each an explanatory caption.

You may be able to handle much of this task on the computer—including formatting the report, generating a first draft, and revising the first draft. You can also enter most of the data for tables. For complex schematics and drawings, however, the standard word processing packages of today will not do the job. In such cases, you might simply leave space for manual drafting and labeling at a later stage. (See Unit 5.5, Word Processing and Document Design.)

The Formal Lab Report: A Suggested Format

Title Page. The front sheet of the report should display the exact title of the experiment, not a modification of the title. Also list the course name and number, the date the report is submitted, the instructor's name, and the name(s) of the writer(s).

Introduction. The introduction tells why, in a technical sense, the experiment was performed. The objectives section of the lab manual may suggest this purpose, and the instructor may expand on it in assigning the project.

Typical objectives of lab experiments are to test theoretical principles, to verify component parameters, to examine the performance of specific circuits or systems, to design or modify circuits or systems to perform given functions, and to troubleshoot malfunctioning circuits or systems.

Equipment and Components. List all the equipment and components, and the number of such items, that you used in the experiment. Include any parameters or specifications for equipment or components that are important to the experimental results. Don't include miscellaneous items such as wire, nuts, screws, and similar materials. The information in the equipment section should be specific enough to allow someone else to repeat the experiment.

Procedure Used. Explain how the experiment was approached, what was done, and how it was done. Include schematics, diagrams, flowcharts, and other illustrations. Also include calculations used to derive data, and mention any special requirements imposed by the experiment. The procedures section should provide enough information so that someone else could repeat the procedure and arrive at the same results.

Observations. In the observations section, you should list the experimental results of the procedures described in the previous section. The data should typically be arranged in tables, with the units of the data specified at the top of columns.

List only raw data here, along with any immediately derived values. If calculations are made with these data in order to reach conclusions, put these into the next section.

Analysis and Conclusions. Discuss the results of the experiment, draw conclusions from the data obtained, develop curves for the data, and qualify points of doubt or error here. In applicable cases (e.g., when experimental results differ widely from expected results), you may wish to suggest modifications in the experimental procedure.

Depending on the nature of the experiment and on its purpose, you may also write separate analysis and conclusions sections.

Example of a Formal Lab Report

A formal lab report for an academic setting is shown in Figure 2–11.

Current and Electromagnetic Force

Physics 201 Laboratory

April 14, 1997

Prepared for

Professor H. J. Wilson

by Angela Aquino

FIGURE 2–11
Formal lab report (pp.95–100).

Introduction

When a current-carrying conductor is introduced to a magnetic induction field, a force vector is generated. The force vector can be described by the general equation $F = IdB$, where F is the force vector, I is current through the conductor, d is the length of the conductor operated on by the magnetic induction field, and B is a constant magnetic induction field.

The relationship described by this equation is the basis for calculations used in such applications as steering electrons in a controlled manner in cathode ray tubes and modulating the magnetic force in magnetic levitation systems.

This experiment was conducted to observe the relationship between the conductor current I and the resultant force vector F as a means of testing the validity of the general equation. For the purposes of the experiment, and primarily because of the size and construction of the solenoid used in the apparatus, two values were predetermined:

$$B = 0.0015 \text{ Tesla}$$
$$d = 4 \text{ cm}$$

Table 1 lists the equipment and components used in the experiment.

TABLE 1

Equipment and Components	
One adjustable 0-20 volt DC power supply (min. 1.5 amps current capability)	One strip of aluminum foil (0.03-g mass)
One apparatus frame	One 5-Ω 5-w resistor (R)
One 56-cm length of 12-gauge aluminum wire	One digital voltmeter (Fluke 3750A)

FIGURE 2–11, *continued*

Procedure

First I bent the aluminum wire into a squared U shape with the left side of the U 23 cm long, the trough 5 cm, and the right side 28 cm. Then I bent 3 cm of the end of the left side and 8 cm of the end of the right side at right angles to the sides of the U. Finally, I bent 5 cm of the right-side 8 cm length perpendicular to the plane of the U (see Figure 1).

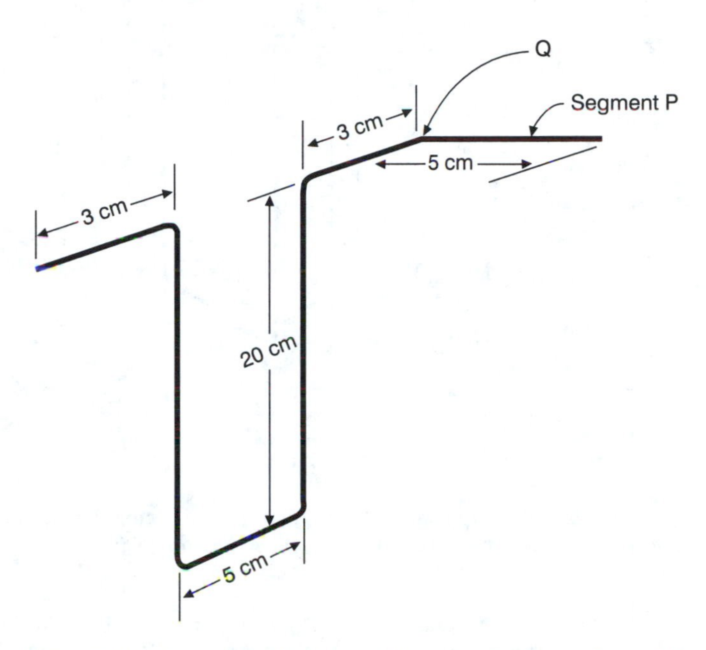

FIGURE 1 Wire bent into U shape, with Segment P perpendicular to plane of U at Point Q.

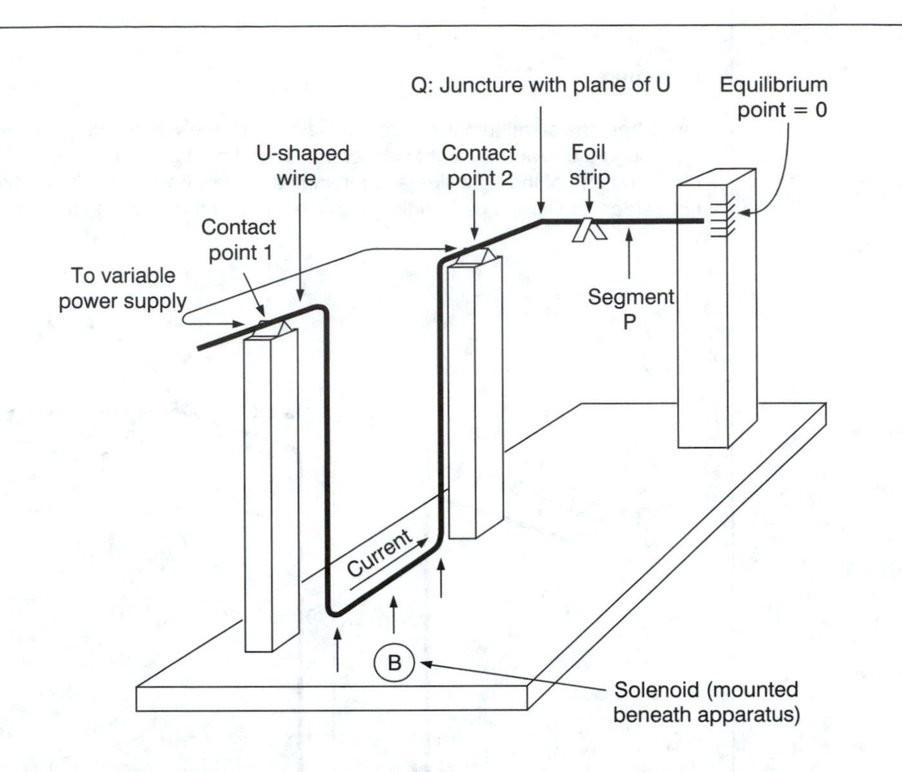

FIGURE 2 Apparatus for measuring force vector on a current-carrying conductor (wire) generated by a constant magnetic induction field.

I suspended the wire on the apparatus frame as shown in Figure 2. A solenoid connected to a circuit to generate the constant magnetic field B was mounted underneath the apparatus. The strip of aluminum foil was initially placed about halfway across the projecting end of the wire pointer (Segment P).

With the apparatus in place, I connected the power supplies to the contact points and to the solenoid, as shown in Figure 3. Then, I introduced test currents through the U-shaped wire via the

FIGURE 2–11, *continued*

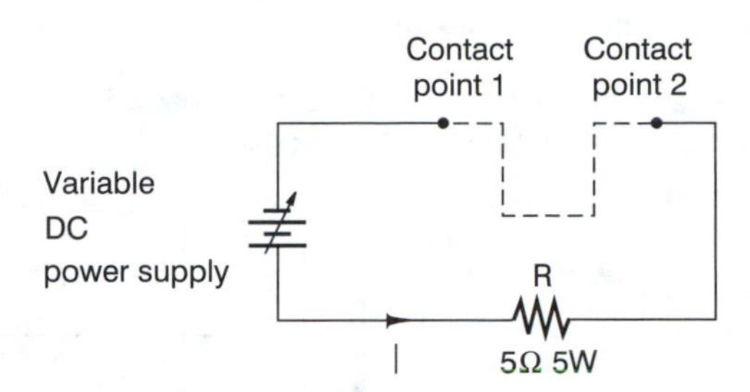

FIGURE 3 Electrical connections for conductor wire.

contact points and moved the foil strip until Segment P was in equilibrium, pointing to zero on the marker bar. For each test voltage, when Segment P was in equilibrium, I recorded the distance from the near edge of the foil strip to Point Q. These voltages and distances appear in Table 2.

Then, for each voltage and distance, I calculated the magnitude of the force vector (F) exerted on the wire. For this purpose, I used the basic equivalence between F and mass x acceleration, combined with the torque effects created by the shape of the wire. This relationship was expressed by the following equation:

$$F = \frac{Mgd_1}{d_2}$$

where M = mass of the foil strip, g = acceleration due to gravity, d_1 = distance from the foil strip to Point Q, and d_2 = 20 cm (the length of the vertical part of the U-shaped wire). Observed values for F appear in Table 2 along with the expected values derived from the general equation $F = IdB$.

Observations

TABLE 2 Expected and observed values of F for a constant magnetic induction field (0.0015 Tesla).

I (amps)	*F: Expected* (Newtons x 10^{-3})	d_1 (distance to Q, cm)	*F: Observed* (Newtons x 10^{-3})
0.25	0.015	1	0.022
0.50	0.03	2	0.037
1.0	0.06	4	0.067

Analysis and Conclusions

The relationship between I and F for the expected and observed cases was extrapolated from Table 2 and plotted in the graph of Figure 4.

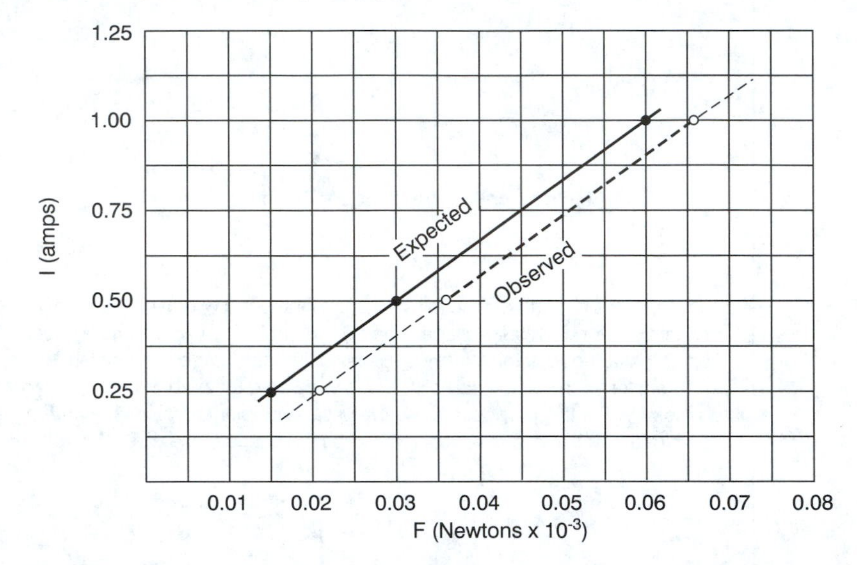

FIGURE 4 Expected and observed relationships between current (*I*) and force vector (*F*).

As Figure 4 shows, the observed relationship between I and F is linear, thus satisfying the general equation $F = IdB$ given in the introduction. Because the observed and expected results are close and the slope of both plots is the same, we can conclude that the difference is a constant that may be a result of the B value generated by the solenoid. This value may not have been exactly 0.0015 Tesla. Contributing factors may have been errors in setting the test currents or in estimating the equilibrium level of the wire pointer or the distance of the foil strip from Point Q.

To avoid incorrect B values, a flux meter could be used and the current I precisely adjusted to create a magnetic field of 0.0015 Tesla at the trough of the U-shaped wire.

FIGURE 2–11, *concluded*

UNIT 2.6 FEASIBILITY REPORTS

Companies are naturally reluctant to invest large sums of money in unproven equipment or new practices, especially when these changes might alter their established methods of doing business. A feasibility study is assigned to investigate the advantages of change given the budgetary and technical constraints of ongoing operations. The study should lead to a sound investigation and a clear report on its findings.

Doing the study and preparing the report will challenge your research skills, technical insight, and organizational abilities. The assignment tests how thorough you are and how well you understand the various factors that impact business decisions. A feasibility report is a good exercise of basic analytical skills and of their application to business strategies.

POSSIBLE ASSIGNMENTS

Write a feasibility report on one of the following:

- ❑ Using cellular mobile phones in dispatching field service representatives
- ❑ Installing fax machines in the branch offices of a technical services company
- ❑ Relocating an organization
- ❑ Selecting a site for a new manufacturing facility
- ❑ Updating and expanding the telecommunications services at a university with an associated teaching hospital
- ❑ Substituting more advanced components for those currently used in a company's product
- ❑ Establishing a program for student purchase of personal computers at a technical college
- ❑ Automating a college library

Approximate length: 5 to 10 pages.

Explanation of a Feasibility Report

Something *feasible* is not only possible but also reasonable and advisable. A feasibility study is assigned by a company to determine whether a possible course of action is also the best course of action for that company. The success of the study depends on how well all the alternatives are researched, including the alternative of taking no action at all (see Figure 2–12).

If you are asked to conduct a feasibility study, your first step should be to clarify the specific purpose of the project. Are you trying to determine whether the company should do something or not, whether it should act now or later, whether to

Alternatives

A_1: Buy new machine X → Cost → Productivity increase
A_2: Buy new machine Y → Cost → Productivity increase
A_3: Repair old machine → Cost → Productivity increase
A_4: Reorganize production → Cost → Productivity increase
A_5: Do nothing → Cost. . . .

FIGURE 2–12
At the heart of the feasibility study is an examination of alternatives and their consequences. One alternative is to do nothing, which has its costs too.

take one course of action as opposed to another, who should act, or even whether there is enough information available to decide on a course of action?

You should be able to state the purpose of your study in a single sentence. For example: "The purpose of this study is to determine whether Bateson Manufacturing should purchase new grinding machines for the Automotive Subassemblies Division at the Hawthorne Works." This statement will provide the basis for the report you write.

Next you should think about the context and limits of the study. How extensive should your study be? What factors will affect the company's decision, and how will that decision be made? What are all the practical alternatives available to the company?

In the case of a major purchase, the company would want to know why it should spend the money. Would the new equipment increase productivity? Would the purchase be cheaper than repairing old equipment? Would the new equipment save production time? The better you understand what is important to your company, the better you will be able to decide what to investigate and include in your report and the more valuable that report will become.

After you have clarified the purpose and scope of your study, you can begin the research. You might do formal library research, call on vendors or experts in a field, analyze company records, or study current practices or procedures. Your findings should be carefully documented for use in the report.

Some of the results of your investigations will bear directly on the central questions you are exploring. For example, costs of purchase and costs of operation of equipment can be compared in objective fashion. Other advantages may need to be inferred or estimated. Relocating the company headquarters or selecting a site for a new retail outlet are examples of choices filled with subjective advantages and disadvantages. You will need to include these intangible considerations as well.

When you have gathered all the information and conducted the analysis necessary to make a decision, you must decide what *you* would do if you had to act for the company. Include this decision in your report as a recommendation.

The four main parts of a feasibility report follow.

1. The introduction, which states the purpose and the conditions of the study
2. The body, which presents your research findings and your analysis of them
3. The conclusion, which summarizes the facts and the analysis
4. The recommendation, which presents your own decision about a course of action

The conclusion and recommendation can also be placed at the beginning of the report to serve as a form of executive summary.

Tables are especially valuable in situations where many factors must be organized and brought together to point toward a conclusion. These tables should be numbered for easy reference and titled to reflect the reason for bringing particular facts together. Figures such as diagrams, graphs, and charts are also useful devices.

A title page, table of contents, abstract, list of illustrations, appendices, and a glossary can also be included. If the company does not specify a format that requires these sections, use your own judgment: A table of contents may be helpful if your report is long and complex but unnecessary if the report is relatively brief.

Example of a Feasibility Report

Figure 2–13 shows a feasibility report for a company considering replacing its copier.

FEASIBILITY REPORT:
REPLACING THE IBM 1403 COPIER

Roger Bernstein
Technical Services Division

The purpose of this study is to determine whether Bateson Manufacturing Company should replace its copier to meet projected increases in printing volumes.

This study looks at the projected copying volumes for the company, alternate copiers as compared to the present equipment, operator times and costs, and special features of copiers.

Volume Projections

Copying volumes have increased 29% in the past year, from 140,000 pages to 180,00 pages per month. New contracts will increase pages-per-month demand to 200,000 by the end of this year. Projections based on new contracts and market surveys indicate a growth in copying volume to 300,000 pages per month by the end of the next 3 years.

FIGURE 2–13
Feasibility report (pp. 103–106).

Current Copier Capacity and Status

The present copier, an IBM 1403, has a capacity of 150,000 pages per month when operated 8 hours per day.

The machine is 3 years old and has needed no major repairs in that time. A maintenance fee that covers all repair costs is paid monthly. Estimated machine life is 6 years. Surveys of other users have supported this estimate:

1. Grosbeck Inc.—This company has operated the IBM 1403 on a full-time basis for the past 4 years with no major repairs.
2. Quay International—This company recently replaced their IBM 1403 after 6 years of full-time service with one major repair.
3. Quinby and Associates—The IBM 1403 at this company has been operated full time for 4½ years with no major repairs needed.

Running the IBM 1403 on an overtime basis would proportionally lower the estimated machine life. Running 200,000 pages per month would add overtime usage that would decrease the approximately 3 years of remaining machine life to 2¼ years.

Furthermore, the IBM 1403 copier has been discontinued and cannot be replaced with the same model.

Operator Overtime

Operator salary rates are currently $7.39 per hour, with overtime paid at time and a half, or $11.09 per hour. Operators currently work about 2 hours of overtime a day to meet copying demand of 180,000 pages per month. Overtime will increase to more than 2½ hours a day by the end of the year, when the new contracts increase volume to 200,000 pages per month.

Other Copiers

Two alternate copiers were investigated as replacements for the IBM 1403. They are the IBM 3211 and the Xerox 8700. These machines represent improvements in duplicator technology within the medium price range. They were selected by the Bateson purchasing department after industry surveys.

Both of the alternate copiers are capable of 200,000 pages per month without operator overtime. A comparison of the monthly basic operating costs of all three machines is presented in Table 1.

Paper Costs

The two IBM copiers use pin-fed, fan-fold computer paper at a cost of $9.10 per thousand sheets for the IBM 1403 and $10.30 per thousand sheets for the IBM 3211. The Xerox 8700 can use any standard 8½-by-11-inch precut paper at a cost of $5.00 per thousand sheets.

The Xerox copier can also print on both sides of the paper; the IBM copiers cannot. Ninety percent of the company's duplicating can be done using both sides of the paper, reducing paper usage to 550 sheets on the Xerox for every one thousand sheets on the IBM copiers.This ratio is reflected in the paper costs in Table 1.

FIGURE 2–13, ***continued***

TABLE 1 Monthly operating costs at 200,000 pages per month.

ITEM	IBM 1403 ($)	IBM 3211 ($)	Xerox 8700 ($)
Lease cost	568	1,108	3,733
Paper	1,820	2,060	550
Meter charge	None	None	2,100
Maintenance	1,187	1,269	0*
Tape drive	418	344	0**
Operator salary	1,721	1,686	1,084
Operator overtime	860	—	—
Sales tax	39	0*	0*
Totals	6,613	6,467	6,972

*Included in lease cost.

**Not required at this volume.

Paper costs have risen 18% over the summer and are predicted to rise by 25% by year's end. With a 25% increase in paper costs inserted into Table 1, the new monthly operation costs would be as follows:

IBM 1403	*IBM 3211*	*Xerox 8700*
$7,068	$6,982	$7,110

Indications are that paper costs will continue to rise. Another increase of 25%, which could occur as early as the first quarter of next year, would make the monthly cost of operating either IBM copier higher than the Xerox.

Special Features

When equipped with a logo font, the Xerox 8700 is capable of printing specially designed forms with the company logo. The font retails for $1,000.

Neither IBM machine has the capability to print the company logo. Current practice is to use outside contractors for these forms. Last month's charges for forms with the company's logo were $235, and demand for these forms has been increasing.

Spreading the cost of the logo font over the 5-year estimated life span of the Xerox 8700 would bring this cost to approximately $16.70 a month. Factoring in this amount, plus the outside charges for the IBMs, brings monthly costs to the following levels:

IBM 1403	*IBM 3211*	*Xerox 8700*
$7,303	$7,217	$7,127

Operator Time

Table 2 shows the number of operator hours required to produce 200,000 and 300,000 pages per month on the three copiers.

With operator time down to 5 hours a day on the Xerox 8700 at 200,000 pages per month, a full-time oeprator could spend the remaining 3 hours each day inputting data. Present workloads for input operators already include overtime.

Table 2 also shows that only the Xerox 8700 would avoid operator overtime at the likely 300,000 pages per month output volume.

TABLE 2 Daily operator time required.

ITEM	IBM 1403	IBM 3211	Xerox 8700
For 200,000 pages/month	10.6 h	7.8 h	5 h
For 300,000 pages/month	15.9 h	11.7 h	7.5 h

Conclusions and Recommendation

The Xerox 8700 is lowest in operating costs when the factors of paper cost increases and logo form printing are included in the totals. The Xerox 8700 has the additional advantage of higher printing speed. Faster printing would free operators to help reduce overtime.

I recommend the company lease the Xerox 8700 beginning with the next fiscal quarter.

FIGURE 2–13, *concluded*

UNIT 2.7 PROPOSALS

The written proposal is a common planning and evaluation tool in organizations facing problems or looking for opportunities. By requiring systematic study of an existing situation and of actions that might create an improvement, the proposal pulls together many of the facts needed to make a decision. And because it usually travels up the chain of command, a well-prepared proposal can bring recognition and advancement to the author. Even if your proposal is not accepted, the work you put into it is likely to pay off in a clearer understanding of the situation you have studied.

POSSIBLE ASSIGNMENTS

Write an informal proposal on one of the following:

- ❑ College administrative arrangements or student services such as registration procedures, bookstore or library policies, or graduate placement services
- ❑ Class or lab arrangements such as field trips, lab-lecture coordination, equipment checkout, or scheduling and design of projects

Write a formal proposal on one of the following:

- ❑ Major design projects
- ❑ System improvements for small businesses
- ❑ Departmental reorganization
- ❑ New product ventures.

Approximate length: 1 to 5 pages for informal proposals; 5 to 25 pages for formal ones. (*Note:* Albert Einstein proposed the building of the atom bomb in a one-page letter to President Roosevelt. The proposal to develop the F-15 fighter plane ran to 30,000 pages.)

Explanation of a Proposal

Spontaneous proposals are made all the time. People meet to talk about common tasks and someone says, "Why don't we . . . ?" or "Why don't you . . . ?" Often, these casual suggestions remain undeveloped because a major flaw in the proposal is quickly apparent.

Yet, the same "Why don't we . . . ?" impulse has been the starting point for many successful projects and occasionally for historic breakthroughs. The impulse comes from the perception of a problem or the more general feeling that things are not as good as they could be. Stated in a more positive way, it comes from recognizing an opportunity.

The written proposal is a full description of a problem and its solution. It gives enough details to allow the reader to evaluate the merits of the proposed action. It also attempts to *sell* an idea and therefore uses the strategies of any effective sales message; in particular, it offers clear-cut benefits and helps the reader to act on the proposal.

Proposals may also be classified as internal or external to an organization. An internal proposal is addressed to a higher level within the organization and suggests changes and improvements. The external proposal is written on behalf of the organization and represents an offer to another organization to solve a problem in a particular way for a specified compensation.

Most internal proposals aimed below the executive management level are categorized as *informal* submissions; internal proposals to senior management and all external proposals are presented as *formal* documents.

The Informal Proposal

Suggested changes in departmental or company procedures, modifications of products or equipment, alternate production methods, new product ideas, staff changes or additions—all these are suitable for handling by means of an informal proposal. In addition, many recommendations that are expected to require further study before final approval is given may be introduced through an informal proposal. In these cases, the informal proposal may lead to in-depth studies and testing that finally produce a full-dress formal proposal.

The designation *informal,* however, should not be taken to mean that the proposal is made casually or that it is poorly thought out or incomplete. The informality applies more to form than to content. The informal proposal pursues the same strategies and contains all the major elements of the formal version but without some of the supporting apparatus such as a table of contents, appendix, or bibliography. A separate sheet or title page may be used, or the informal proposal may be totally incorporated into a letter or memo.

A model of the informal proposal (after the title page) follows.

1. Introduction
   a. Problem
   b. Solution
2. Analysis
   a. Background
   b. Causes of the problem
   c. Scope, significance, implications of the problem
3. Detailed solution
   a. Work and management: people, materials, times, costs
   b. Drawbacks
   c. Benefits
4. Action stimulus

The Introduction. Without using background information or a lead-in, the *introduction* directly states the problem and its proposed solution. These statements may be as brief as one sentence each. Their purpose is simply to orient the reader to a subject that will be discussed more fully in the following sections. Giving the solution at the very beginning keeps the reader focused on a particular proposal instead of thinking about alternatives.

The Analysis. The *analysis* of the situation or problem is a key section that should be carefully thought out before the proposal is even started. Maximum effort should be given to determining the causes of the problem so that the proposed solution is aimed at eliminating these causes and not merely treating symptoms of the problem (see Figure 2–14). For example, if I repeatedly find a small puddle of radiator fluid under my car, my initial (and temporary) solution might be to buy some sealant to plug the leak in my radiator. However, closer inspection shows that the radiator cap is worn down, allowing the fluid to boil up and escape via the overflow tube. If I treated the symptom by my first solution, the problem would continue; by determining the actual cause and buying a new radiator cap, I eliminate the problem.

The analysis section should also include a brief account of the background of the problem—what led up to it, how long it has existed, and similar information—so that readers who are not close to the situation can better understand it. After sketch-

FIGURE 2–14
The challenge in your analysis of the problem is finding the cause, which is often obscured by its symptoms.

ing the background and identifying the causes, the writer should discuss the significance of the problem by pointing out the cost (in time, money, efficiency, and morale, for example) of allowing the problem to continue. This part of the analysis can motivate readers to take the proposal seriously and to act on it more quickly.

The Detailed Solution. The *detailed solution* explains the proposal and how it can be accomplished. The explanation should be accompanied by necessary diagrams, plans, flowcharts, and other graphic aids. Explaining "how it can be accomplished" requires a listing of the people and materials that would be needed and the timing of the proposed change. If lengthy development or phasing-in is suggested, a schedule should be presented.

To make the proposal convincing, you should follow two basic strategies. On the one hand, your proposal should be realistic; on the other, it should show the clear advantages of the change. Rather than glossing over the drawbacks of the proposal, you should bring them up yourself so they can be balanced against the advantages. A proposal that ignores drawbacks allows the readers to think of drawbacks themselves and creates the impression that the proposal is unrealistic. A proposal is realistic and convincing when it itemizes costs and other drawbacks but offers clear benefits that outweigh the negative factors.

The Action Stimulus. The proposal should end with a suggestion that helps the reader take *action.* The essential message is, "Please contact me after you've read this, and let's discuss it." Listing a phone number or suggesting when and where you are available to talk can help bring a definite response.

The Formal Proposal

Although the formal proposal follows the same basic strategies as the informal version, there are occasions when a formal presentation is preferred. These include proposals to senior management, outside organizations or other companies, and those that involve complex projects, lengthy time periods, and large sums of money.

Frequently, government agencies and other organizations that receive many proposals ask that a particular organizational scheme and particular format be followed. In such cases, the specified method of organizing the proposal serves the convenience of the reviewers and allows them to better compare competing proposals.

In the absence of a specified format for a formal proposal, use the following:

1. Letter of transmittal (or memo)
2. Cover sheet or title page
3. Summary or abstract
4. Table of contents
5. Body of proposal
    - **a.** Statement of problem
    - **b.** Proposed product, system, or methodology
    - **c.** Facilities and equipment required
    - **d.** Personnel required
    - **e.** Duration or schedule
    - **f.** Costs
    - **g.** Advantages and disadvantages
6. Supporting materials
    - **a.** Bibliography
    - **b.** Appendix

As Figure 2–15 shows, the formal proposal corresponds closely to the informal proposal in the key analytical and solution sections.

The Letter of Transmittal. The letter of transmittal, which may also be in the form of a memo, brings the proposal to the attention of the reviewer and identifies its highlights. It is not a part of the proposal itself, which starts with the next page.

The Title Page. The title page includes a title typically beginning with the words "A Proposal for . . . "; it also identifies the person or organization to whom the proposal is submitted and the person or organization submitting the proposal. A date is always given, and an identifying number (typical for solicited proposals) is located at the top of the page.

The Abstract. An informative abstract is used to present a brief summary (about one-half page) of the entire proposal, including a statement of the problem and the solution. This section is essentially a slightly longer version of the introduction of the informal proposal.

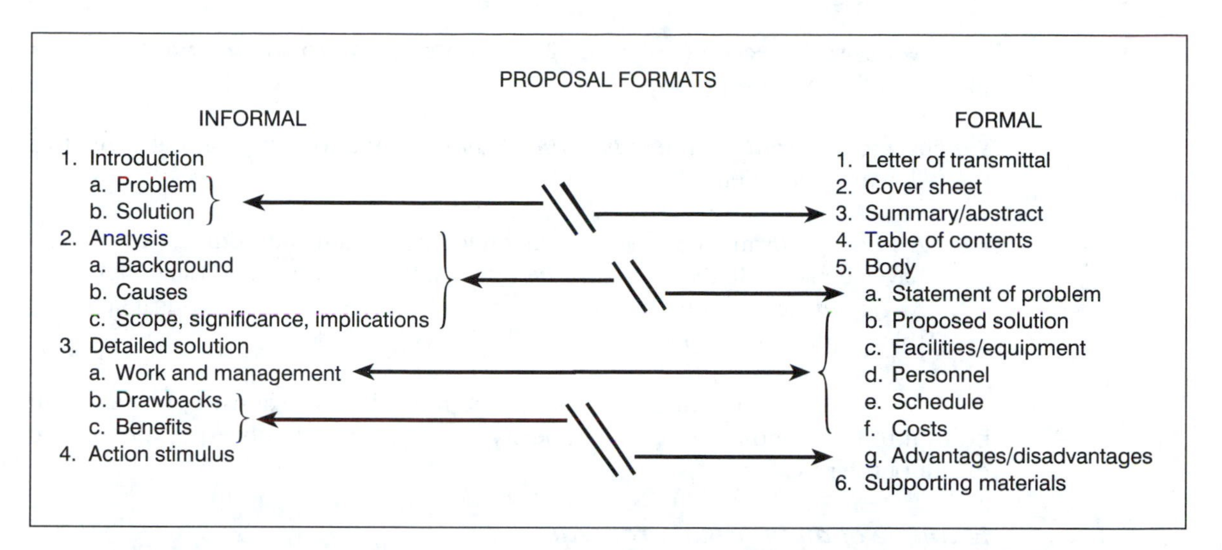

FIGURE 2–15
Formal and informal proposal strategies compared. The key analytical and solution sections of both proposal types are the same.

The Table of Contents. The table of contents gives the page numbers of major sections and subsections.

The Body. The body of the proposal includes the following.

1. A statement of the *problem* in full, including its background, causes, and significance. This section corresponds to the analysis of the informal proposal.
2. An explanation of the proposed *product, system,* or *procedure* for solving the problem. This should be a detailed, precise technical description with exact specifications and supporting diagrams and charts.
3. A description of *facilities,* such as laboratories or plants, and *equipment,* such as test instruments, construction machinery, or computer hardware.
4. A listing of the *personnel* who will be needed to do the job. If these people are not known to the proposal reviewers, resumes or professional biographies should be attached.
5. A statement of the *timing* of the project, including its entire duration and a schedule for the accomplishment of major subparts.
6. An itemized breakdown of the *costs* of the proposal according to categories such as salaries, capital equipment, and overhead.
7. A listing of the *advantages and disadvantages* of the proposal. Both elements are important to making the proposal realistic and convincing. A comparison of the proposal to alternative solutions can help show why the recommended approach is best.

Note: The preceding Sections 2 to 7 correspond to the ***detailed solution*** offered in the informal proposal.

Supporting Materials. The *supporting materials* of a formal proposal consist of the following attachments:

1. A *bibliography* may be attached listing references examined, some of which may have been cited in the proposal (see Unit 5.4, Bibliographies and Citations).
2. An *appendix* may contain full copies of studies or reports or statistical tables that bear on the proposal.

Note: These supporting sections may be more or less extensive, or may even be omitted, depending on the complexity of the proposal and especially on the extent of the research.

Example of an Informal Proposal

An informal proposal for projects in an undergraduate program is shown in Figure 2–16.

A PROPOSAL FOR
FRESHMAN-SENIOR PROJECT TEAMS

to
Dean of Academic Affairs
Madison Institute of Technology

submitted by
Amanda B. Petrakis
Rogers & Hertz Consulting, Inc.
May 8, 1997

FIGURE 2–16
Informal proposal (pp. 112–115).

To many new students, Madison's programs seem long and difficult. The students' inability to see the eventual linkages and career applications of the subjects they study may contribute to the frustrations that cause them to drop out. A way of motivating new students to persist is to put them into project teams managed by seniors working on their design projects before graduation.

Analysis

Recruiting presentations to potential students describe Madison programs as practical and applications oriented and focused on careers. As a result, students arrive at Madison expecting to be quickly moving toward career competence.

Despite the attempt to give freshman students applications in labs and elsewhere, they must also master fundamental concepts and techniques in many different areas. This may be a logical way to design a program, but it may also create the impression that subjects are unrelated, that there are many of them to be learned, and that the end of the program is a long way off.

Most of the students who leave Madison do so within the first three terms, and the highest drop-out rate occurs in the first term. Those who leave cite a variety of reasons, but the factor that seems most directly related is academic performance. Other factors being equal, those who do well in their courses stay; those who do poorly leave. Further, the academic problems of those who leave seem to have a significant motivational basis.

The consequences of a high drop-out rate in the early terms are severe for both Madison and its students. To the college, the loss is not only a short-term loss of tuition revenues. The permanent image of Madison as an institution committed to student success cannot accommodate a continued large failure rate. For the student who drops out, the consequences can be measured in financial losses, time lost in establishing a career, and lowered self-esteem and confidence.

A Solution to the Motivation Problem

Although it is not offered as a cure-all, the organization of project teams composed of new students managed by seniors could positively affect student motivation and persistence. This approach also offers benefits to the upper-term students. The ultimate beneficiary of a successful program would be Madison Institute of Technology.

How the Program Would Work

In the first week of the term, seniors enrolled in capstone project courses sign a list indicating that they wish to manage a group of freshman students. The groups would help them complete their senior projects. With the number of teams thus established, each prospective manager prepares a one-sentence description of his project by the beginning of the second week. These descriptions are compiled into a single list.

Meanwhile, new students are acquainted with the program during orientation classes in the second week. In the third week, the list of project descriptions is circulated through the class, and students sign their name (and phone number) by the project of their choice. The maximum number of team members has been set somewhere between three and ten, depending on the number of projects and the number of freshmen in the program. Any discrepancies between the assignments desired and the number and types of projects available are adjusted through consultation between the instructors of the freshman and senior courses.

With the team assignments set, the project managers contact their team members. They arrange a meeting to explain the project more fully and to assign duties. The tasks assigned would vary according to the project. Some thought should be given to the selection of tasks that are within the capabilities of new students but that are also helpful to the senior student. Possible activities include the following:

- ☐ Library research
- ☐ Writing to, phoning, or visiting parts stores, software shops, or local businesses
- ☐ Word processing and graphics design assistance
- ☐ Soldering, assembling, testing
- ☐ Record keeping
- ☐ Report preparation assistance

A project schedule is prepared. It shows milestones for the completion of subparts and a schedule of regular team meetings to review progress. The project manager should meet with the group at least once a week. He should provide a brief written evaluation of each team member's work at the end of the term. Copies go to the student and the orientation course instructor. The final project report should also include an evaluation of the effectiveness of the team approach.

A schedule of the implementation stages of the project-management program is shown in Table 1.

Benefits and Drawbacks

In a program of this size, the sheer number of students involved would create a need for coordination and monitoring. Because of the inexperience of the freshmen, mistakes in the work would be made; because of the seniors' inexperience in project management, human errors would also occur. The possibility of frustrations, headaches, and hard feelings among the participants thus needs to be weighed against the potential advantages of the program.

The fact that mistakes are likely to be made within an academic program may itself be seen as a learning experience. In their supervision of the program, faculty can limit the impact of mistakes and point out ways of avoiding future problems. Mistakes under these conditions are also preferred to mistakes made after graduation, when they may affect the employer's bottom line.

To the freshmen, the program offers a realistic glimpse of their intended career field, both in terms of subject matter and in working as part of a team. By carrying out tasks within their capabilities, freshmen gain a sense of accomplishment. Their contacts with a senior student present them with a role model—of someone who has succeeded in the academic program they are just beginning. By working with their project leader and team members, they receive emotional, social, and academic support, all of which should help their morale in the critical first term.

For seniors, the program offers practical help in completing a project in a busy final term. It also gives them a chance to apply the interpersonal, managerial, and

FIGURE 2–16, *continued*

TABLE 1 Schedule for freshman-senior projects.

Weeks	Seniors	Freshmen	Administration
1	Sign up for project management		
2	Submit 1-sentence description of project	Learn about program in orientation class	Compile list of projects, set maximum team size
3	Contact team members	Sign up for project teams	Adjust any discrepancies
4	Meet team members, set duties	Meet project managers	
5-12		Work on projects, meet weekly to review progress	Monitor program through project and orientation instructors
13	Write evaluations of team members	Evaluate project experience in orientation classes	
14	Submit final project report		Discuss program in orientation class

planning skills they have been acquiring in their academic coursework. The experience of managing a project team should gain favorable attention from employers and should more than offset the difficulties involved.

The time and effort required from Madison faculty and administrators thus promises to be well rewarded in terms of student success. By bringing freshmen and senior students together, the college can increase retention levels and send better-seasoned graduates into their careers.

A Suggested Response

Understandably, a decision about a program of this size requires consultation among faculty and staff. If the proposal is felt to have merit, I would be glad to discuss it further with all interested parties and to develop more detailed plans. I can be contacted at Rogers & Hertz at x363.

Unit 2.7 Case

You work part-time as a student assistant in your college library, which is located in an older building that has been filled to capacity by print and electronic resources and has become crowded and noisy as a result. The only available space in the building is a smokers' lounge adjoining one end of the main reading room. When you suggest to the head librarian, Dr. Marion McArdle, that a wall could be torn down and the smokers' space converted to library uses, she says she only wishes it could.

"I've told Dean Dombrowski we have a critical need for space, and that conversion of this lounge would give us an additional 900 square feet for 24 new study carrels, twelve 300-book shelves, and 8 display racks. The dean has been sympathetic, but says his hands are tied. Some provisions have to be made for the smokers, he says, evidently because of a policy adopted by the board of trustees. This seems to be the only available space at this end of campus."

"Why couldn't some ash trays be placed around the back doors of the building and smokers be required to go outside to practice their habits?" you ask.

"Maybe something could be proposed along those lines," Dr. McArdle replies. "But it would require an awning of some sort for inclement weather, and some provisions would have to be made to police the cigarette butts and other litter, despite the ash trays."

"This is a matter of priorities," you say. "We ought to be able to engineer a solution for something so badly needed by the college."

"You may be right. Why don't you draw up an informal proposal for the project? Address it to me, and I'll forward it to Dean Dombrowski with my supporting comments."

Prepare an outline of the proposal you would submit in this case. Do some analysis of costs and benefits. Write your proposal to Dr. McArdle.

UNIT 2.8 USER'S MANUALS

Even as a new employee, you may be asked to help develop a user's manual for a product. This is extremely valuable experience because you not only learn more about the product but also about strategic matters such as the company's design concepts and service policies. This assignment will prepare you to contribute more quickly to your employer's needs in these cases.

The assignment also challenges you to apply skills that you have been building throughout your academic career. It asks that you incorporate and organize several different writing techniques in consideration of the user's needs. Perhaps in no other kind of professional writing is audience awareness as important.

POSSIBLE ASSIGNMENTS

For a shorter assignment that addresses only certain sections of the user's manual, complete one of the following:

- ❑ Prepare a set of warnings to the user against misuse of a typical consumer product.
- ❑ Prepare two sets of operating instructions for a digital multimeter, one set for a working technician and one for a new electronics student.
- ❑ Prepare a set of maintenance instructions for a familiar piece of equipment, including how often different procedures should be performed.
- ❑ Prepare a list of 10 technical procedures that a nontechnical user might perform, and then write instructions for performing one of the procedures.

- ❑ Prepare a three-page report that includes only certain sections of the user's manual, such as the introduction and warnings, instructions on part of a procedure, or maintenance information only.

For a full-length assignment, either individually or in a team, prepare a complete user's manual for one of the following:

- ❑ Consumer product such as a vacuum cleaner or lawn mower
- ❑ Telephone answering machine
- ❑ Cellular telephone
- ❑ VCR
- ❑ Personal computer
- ❑ Software package
- ❑ Computer peripheral
- ❑ Piece of test equipment
- ❑ Synthesized signal generator

Approximate length: From 1 to 12 pages as an assignment; in professional practice, manuals can exceed 100 pages.

Explanation of a User's Manual

A microcomputer is sold with instructions for setting up and using the product. A bottle of aspirin is capped with instructions on how to open it and is labeled with advice on dosages. Almost every product on the market comes with some type of guidance for the user.

Such messages are often a legal necessity. Consumers who aren't warned against the potential hazards of using a product may seek damages from a manufacturer if they come to harm. Conversely, in the absence of a licensing agreement, users may copy and distribute software in unintended ways. And government and the military typically require user information on the products they purchase.

Manufacturers also benefit from providing written information to the user. If customers understand how to use and maintain a product properly, they will obtain the most satisfaction from it. They are more likely then to become repeat customers. A properly used and maintained product also means the manufacturer will have to spend less time and money servicing it or dealing with complaints. Organizations, too, may be sold on equipment that comes with effective instructions, because these can reduce training time for employees.

The need for clear, helpful user instructions rises in proportion to the complexity of the product. The length of these instructions will necessarily increase too. For complex products, such instructions take the form of a user's *manual.*

The user's manual can be aimed at a technical or nontechnical audience. Some manuals are for service personnel, for example, while others are for nontechnical consumers. Many manuals combine material directed to both audiences.

In any case, the target audience must be identified before a user's manual can be prepared. Then, with the reader in mind, you can decide on a plan of organization for the manual.

Following is a good general-purpose plan for a user's manual.

1. Cover
2. Table of contents
3. Introduction
   a. General technical description
   b. Instructions for using the manual
   c. Important user information (warnings, cautions, notes)
4. Installation or assembly instructions
5. Operating instructions or explanations
6. Troubleshooting information
7. Maintenance information
8. Service information (if applicable)
9. Parts list (if applicable)

Let's look at these sections more closely.

The *cover* of a user's manual should identify the product by name and model or part number, if any, and give the manufacturer's name. The cover can also include a photograph or line drawing of the product.

The *table of contents* should list the titles and page numbers of all major sections.

The *introduction* can be used to accomplish three purposes: (1) generally orient the reader by providing an overall technical description of the product; (2) orient the reader to the manual's organization; and (3) provide important warnings, cautions, and other information the reader should know before attempting to use the product (see Figure 2–17).

Installation or *assembly instructions* are naturally placed before operating instructions. These instructions tell the reader how to put the product into service. They can range from how to unpack the product to how to modify it for a specific application. If the product is typically used just as shipped, you might instead provide custom installation instructions as an appendix to the manual.

Every step in these and the following instructions should be detailed to suit the reader. Never assume the reader will execute a process naturally. For example, don't assume the nontechnical reader will use the correct set of bolts to attach a panel to a chassis unless told to do so in your instructions. Also, if some procedures must be repeated at different points in the process, repeat the instructions each time rather than referring the reader back to the earlier instance. The reader may try to complete the procedure from memory rather than go to the trouble of looking back.

A large number of illustrations are often included with the instructions in a user's manual. Certain procedures can be most clearly described through an illustration. These illustrations, however, should supplement, not replace, the verbal instructions.

Operating instructions for the product would come next for all products that have specific functions and applications and that require user interaction (such as

HEWLETT-PACKARD MAKES NO WARRANTY OF ANY KIND WITH REGARD TO THIS MATERIAL, INCLUDING, BUT NOT LIMITED TO, THE IMPLIED WARRANTIES OF MERCHANTABILITY AND FITNESS FOR A PARTICULAR PURPOSE.

Hewlett-Packard shall not be liable for errors contained herein or for incidental consequential damages in connection with the furnishing, performance, or use of this material.

This document contains proprietary information which is protected by copyright. All rights are reserved. No part of this document may be photocopied, reproduced, or translated to another language without the prior written consent of Hewlett-Packard Company.

The information contained in this document is subject to change without notice.

FIGURE 2–17

A warranty disclaimer statement (From 1992 *HP LaserJet 4* user's manual. Reprinted with permission of Hewlett-Packard Company).

computers, mechanisms, devices, etc.). An *explanation of the operation* of the product would be provided instead when the product has many potential functions that depend on the user's own applications. For instance, a computer designer would need to know the capabilities of an integrated circuit to decide if the circuit would be useful in a design.

Troubleshooting information is provided for minor product faults or common errors that occur in assembling, installing, or operating the product. These errors are usually discovered when the product is tested before being sold. This information is often provided in chart form with problems listed in one column and possible solutions in another.

Maintenance information can be provided for nontechnical users to cover simple procedures they could perform safely, such as cleaning the product or replacing parts that get a lot of wear. Information on serious maintenance and service procedures is usually reserved for qualified service personnel. Often maintenance information is given in chart form, with notations such as "every 2 months" or "every 6 months" next to the procedures.

Service information is provided if the manual will be used by both consumers and service technicians. This section of the manual is more technical than others. Yet, to avoid errors and simplify the service, the technician's guide should also be clearly written in a step-by-step manner.

In fact, two main errors are more likely to be made by the technician than by the user. First, the technician's familiarity with the unit may lead to a procedural error from skipping the instructions. Also, the technician is more likely to be injured because of greater exposure to moving parts or voltages inside the machine. Therefore, safety precautions should be highly visible in a service guide.

A *parts list* is provided only if service information is also included. Typically, exploded-view illustrations (i.e., parts separated for viewing) accompany the parts list, with each part identified in reference to the list.

Audience Awareness

In most cases, user's manuals are directed to readers with a limited knowledge of the product and the technology behind it. Your language should be adapted to suit these readers. An example of a proper readability level can be seen in a typical newspaper article on a technical subject. Only universally accepted technical terms should be used, and abbreviations should be carefully defined or avoided. Even a section such as the service information, which is directed to technical readers, should still use a simple writing style.

In all sections of the manual, you should try to anticipate problems or hazards the user might encounter. To forestall accidents, display a warning or caution separate from the main text. Also, identify any steps that must be taken only by a technically trained person. Do not offer instructions to a general user that should be handled by a specially trained individual. This is one area where legal liability becomes an issue.

Anticipate also that the user will follow the instructions improperly or even attempt to use the product in improper applications. One way to help avoid this is to provide checkpoints within your instructions, for example, "At this point, the green light should be on and you should hear the motor running. If not, STOP, and repeat the last three steps."

Example of a User's Manual

The partial example of a user's manual in Figure 2–18 addresses the first-time, nontechnical user.

The manual's cover identifies the product (a fictitious device) and gives its model number.

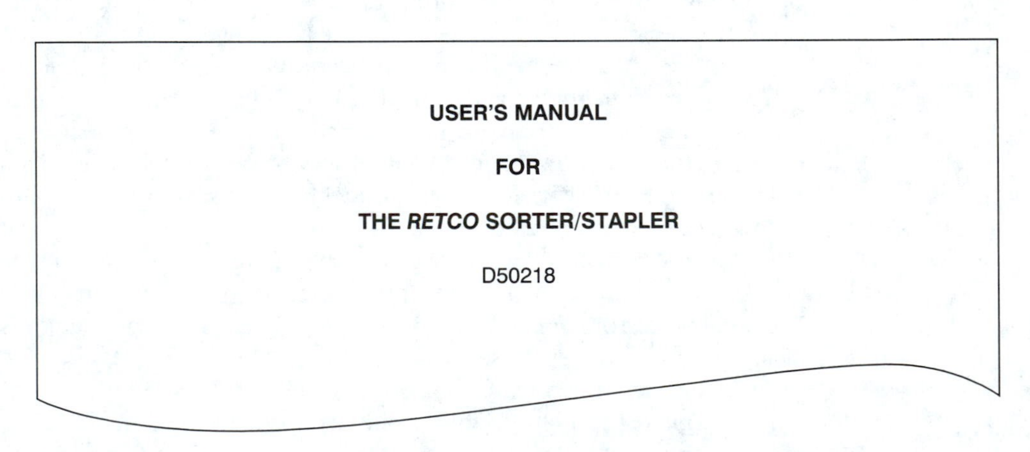

USER'S MANUAL

FOR

THE *RETCO* SORTER/STAPLER

D50218

FIGURE 2–18
Excerpts from a user's manual (pp. 120–125).

The table of contents lists every section and subsection by its page number.

Table of Contents

The introduction begins with a formal technical description of the product. Before the physical description begins, the device is defined, and its functions are explained. By also including the limits of the device, the writer reduces the chance of misapplications. It is common to duplicate this part of the description on a label attached to the device.

INTRODUCTION

Description

The Retco Paper Sorter/Stapler is a vacuum-operated copying machine attachment that will sort copies into groups (from 1 to 15) and then staple and stack each group. The device can accommodate groups of up to 150 pages and can handle paper sizes from 3" x 5" to 11" x 17". It will attach to the following copy machines:

[List of machines here]

The Retco Sorter/Stapler consists of three main parts. These are the load chute, the stapling bin . . .

The introduction continues by describing the manual's plan of organization. Notice how the reader is directed to the Important User Information subsection that contains warnings and cautions.

How to Use This Manual

Before you refer to any section in this manual, and before you attempt to use the Retco Sorter/Stapler, be sure to read the Important User Information that follows this explanation.

This manual is divided into sections for easy reference. The first-time user should read . . .

The introduction then lists the warnings, cautions, and notes on the device for the reader. Notice that the warnings are listed first, with an additional warning to read them. Because a reader still may skip this section, the same warnings are usually duplicated elsewhere within the text. (For instance, the first warning listed here should be duplicated before the first installation instruction.) Often this list is a summary of the warnings throughout the text and is compiled after the rest of the manual is written.

Important User Information

WARNING

Be certain to read all the following warnings before using this device. Personal injury or damage to the device may result if these warnings are not heeded.

****Warnings****

1. This unit is to be installed *only* on the copiers listed in the description section above.
2. Use only Retco staples in the stapler. Damage may result if nonstandard staples are used.
3. Keep hands, clothing, and jewelry clear of the IN and OUT feed chutes. Severe personal . . .

The installation instructions begin with a warning to follow the written instructions. The instructions prepare the reader for the installation procedure by pointing out necessary tools.

INSTALLATION INSTRUCTIONS

****Warning****

Do not attempt to assemble this unit without reading all the instructions that follow. Failure to do so may result in personal injury or damage to the device.

Have the following tools available:
1. Spade clip tool (included in the bag of miscellaneous hardware).
2. Flat-blade screwdriver
3. Pliers

Check to see that all of the following parts were included in your shipment (the letters in the list correspond to those in Figure 5):

A. Load chute
B. Stapling bin
C. Vacuum drive unit
D. Vacuum hose assembly
E. Two (2) mounting plates
F. Power cord
G. Five (5) staple clips
H. Miscellaneous hardware

FIGURE 2–18, ***continued***

After instructing the user on mounting the load chute to the copying machine (not shown), the instructions explain how to attach the stapling bin to the load chute. Notice the warning before the reader begins this procedure and the instruction to keep the manual handy for viewing.

INSTALLATION INSTRUCTIONS (cont.)

17. Tighten the bolts holding the load chute to the mounting plate.
18. Lay the manual on top of the load chute with the manual open to Figure 9.

****Warning****

In the following ten steps, the stapling bin will be attached to the load chute. The stapling bin is heavy and awkward to lift into place. Ask someone to assist you.

19. Lift the stapling bin and position it so the large cutout faces the load chute.
20. *Gently* slide the four (4) pins that project from the stapling bin into the matching holes on the load chute (see Figure 9). *Do not force the pins in.* You may need to wiggle the stapling bin gently until the pins fit.
21. Slide the pins all the way in.

Notice that in Step 21 the writer has taken nothing for granted in the user's performance of the procedure.

INSTALLATION INSTRUCTIONS (cont.)

28. Complete the attachment of the stapling bin by sliding the rubber noise gasket into the slot between the stapling bin and the load chute. The assembly should look as shown in Figure 12. Be sure the two lock tabs are *both* facing the . . .

Step 28 includes two checkpoints for the reader. The first is a visual checkpoint and the second is a safeguard against possible error.

The operating instructions begin with a caution. A caution is used instead of a warning when the potential problem does not involve personal injury or damage to the device. A "notice" is given between steps 5 and 6 as a courtesy to the reader. A notice is used, rather that a caution or a warning, because the notice could be skipped without any adverse results.

Solutions to problems in the troubleshooting information section should flow from the simplest to the most complex. Never advise users to try to fix something they are not qualified to work on.

OPERATING INSTRUCTIONS

****Caution****

Check the hoses running from the vacuum drive unit to the load chute. Straighten any kinks. If a hose is pinched, the load chute will not be able to move the paper.

5. Turn on the copying machine and check that it is operating properly.

Notice: In the following step the load chute will make a loud humming noise. This is normal. The noise will stop after paper is fed into the load chute from the copying machine.

6. Flip the POWER switch on the vacuum drive unit to the "on" position. You will hear the vacuum pump begin to operate. When the green "ready" light comes on, the sorter/stapler is ready to operate.

TROUBLESHOOTING INFORMATION (cont.)

Problem	*Solution*
14 — Ready light fails to come on.	1) Make sure power cord is plugged in. 2) Make sure power is available to outlet. 3) Reset the circuit breaker next to the power switch by pushing it in and holding it for 3 seconds.

IF THESE STEPS FAIL, CALL YOUR LOCAL SERVICE REPRESENTATIVE

FIGURE 2–18, ***continued***

In this maintenance information, instructions for the various procedures are given largely through illustrations for two reasons. First, the procedures are simple and easily explained through pictures. Second, these procedures will be repeated periodically, and illustrations are quick reminders for the user.

MAINTENANCE INFORMATION

Once every 3 months

1. Lubricate the idler wheel inside the load chute (see Figure 35).
2. Change the vacuum line filter as shown in Figure 36.

****Warning****

Failure to provide the above maintenance will void the warranty on this product. It may also cause extreme damage to the device and/or the copying machine to which it is attached.

After explicitly stating that the unqualified user should not attempt to service the device, this section includes a message to the service technician. Because models of this device could change over time, listing the service-guide number for service personnel will help them avoid errors resulting from using the wrong information. This number also appears on the device itself.

SERVICE INFORMATION

Service of this device should be done only by a qualified Retco service technician. There are no user-serviceable parts inside the device except those noted in the maintenance information section.

To Service Personnel:

See the Retco service guide #134-SS.

UNIT 2.9 CASE ANALYSIS

The case method itself is a valued teaching tool, widely used as a focus for class discussion. Cases are also assigned for written analysis. Many business texts include cases at the end of a chapter. So do texts in the social sciences, law, medicine, engineering, and other fields. There are also casebooks presenting longer and more detailed descriptions of business situations.

In business classes, instructors assign cases to test understanding of key concepts and their application to common business problems. A systematic, logical approach to writing a case analysis can give you a clearer grasp of a business situation and help you present solutions more effectively. Working through a case can be a real challenge; it can also provide some of the excitement and rewards that come from solving real-life problems.

POSSIBLE ASSIGNMENTS

Write an analysis of a business case involving one of the following issues:

- ❑ Communication problems in a company or department
- ❑ Management control problems affecting quality of output
- ❑ Financial planning or budgeting challenges
- ❑ Lack of coordination between production, purchasing, and inventory
- ❑ Organizational and staffing problems
- ❑ Obsolescence of employee skills because of advancing technology
- ❑ Marketing issues related to competition
- ❑ Conflicts of interest between departments
- ❑ Leadership and motivation problems

Approximate length: two to five pages.

How to Write a Case Analysis

A business case is typically a description of a situation followed by two or three broad questions or simply a request for an "analysis." In essence, the student is asked to do two basic things: (1) determine the problem and (2) offer a solution. The most productive response therefore is to use the classic problem-solving approach of the manager or consultant. Adapted for case analysis, the process can be summarized as follows:

Mental Work.

1. Determine the problem.
2. List alternative solutions.

3. Evaluate each alternative.
4. Select the best alternative.
5. Decide how to implement the chosen alternative.

...Helps you to get into your writing

This is the analytical process that will give you the best chance of seeing through the case. It also lends itself well to the *writing* that you will need to do. This thought process is described in greater detail in the following section.

1. ***Determine the Problem.*** To find the problem, start by reading the case carefully. After you gain an overall grasp of the situation, jot down some of the key information that may be relevant to your analysis. From these elements, focus your thinking on identifying the problem.

Finding the problem is a logical search for the *cause* of the various *symptoms* described in the case. The cause and its symptoms are distinct and should not be confused with each other. If you eliminate the cause, the symptoms will disappear; but if you attack only the symptoms, the cause of the problem will remain.

Many business cases, for example, describe companies that are losing sales or suffering earnings declines. These financial indicators, however, cannot be the causes of the problem because something else must be causing them. If you look closely at the other information in the case, you will usually find that the problem stems from poor quality, ineffective marketing, personnel conflicts, deficient planning, or similar causes. The "problem" therefore will be something the company is doing or not doing that leads to a number of undesirable results including a shaky bottom line.

2. ***List Alternative Solutions.*** To identify alternatives, start by brainstorming. Write down as many potential remedies as you can think of, without worrying about the soundness of the ideas at first.

Then do some pruning: Some items will be repetitive, others will be clearly unrealistic, still others will miss the target. Throw these out, and reduce your initial list to the most likely remedies.

3. ***Evaluate Each Alternative.*** To evaluate the remaining alternatives, focus on the *consequences* of each remedy. Where possible, estimate the consequences in numerical terms (so many units of production, so many dollars of revenue, etc.). Then compare these projections. (See Figure 2–19 for an example of this technique.)

Some consequences, such as the impact on employee morale or on the company's image, may not be quantifiable. In such cases it is still helpful to clearly identify the likely outcomes and to compare them to other outcomes as the basis for your choice of remedies.

4. ***Select the Best Alternative.*** By looking at consequences, you should be able to make the best choice as the next logical step. When you do, prepare also to *defend* your choice. This will be required in your written analysis. Your "defense" should be based on your comparison of outcomes in the previous step. In Figure 2–19, the outcomes can be stated as expected revenues. The revenues, in turn, are based on the probabilities of various levels of sales of the new PCs.

*5. **Decide How to Implement the Chosen Alternative.*** In any organization, implementation can ensure the success or failure of a decision. People must be informed about the change and prepared for it. A well-planned implementation therefore requires that you decide at least the following:

- ☐ Who should do it?
- ☐ When should it be done (now or later, all at once or phased in, etc.)?
- ☐ What methods and means should be used (an announcement by the president, a series of departmental meetings, etc.)?
- ☐ How will its success be measured? (Come back to your expected outcome and state it in measurable terms.)

Think through the preceding five steps as you prepare to write your analysis of the case. They will give you not only stronger arguments but also a rough outline of the written analysis itself.

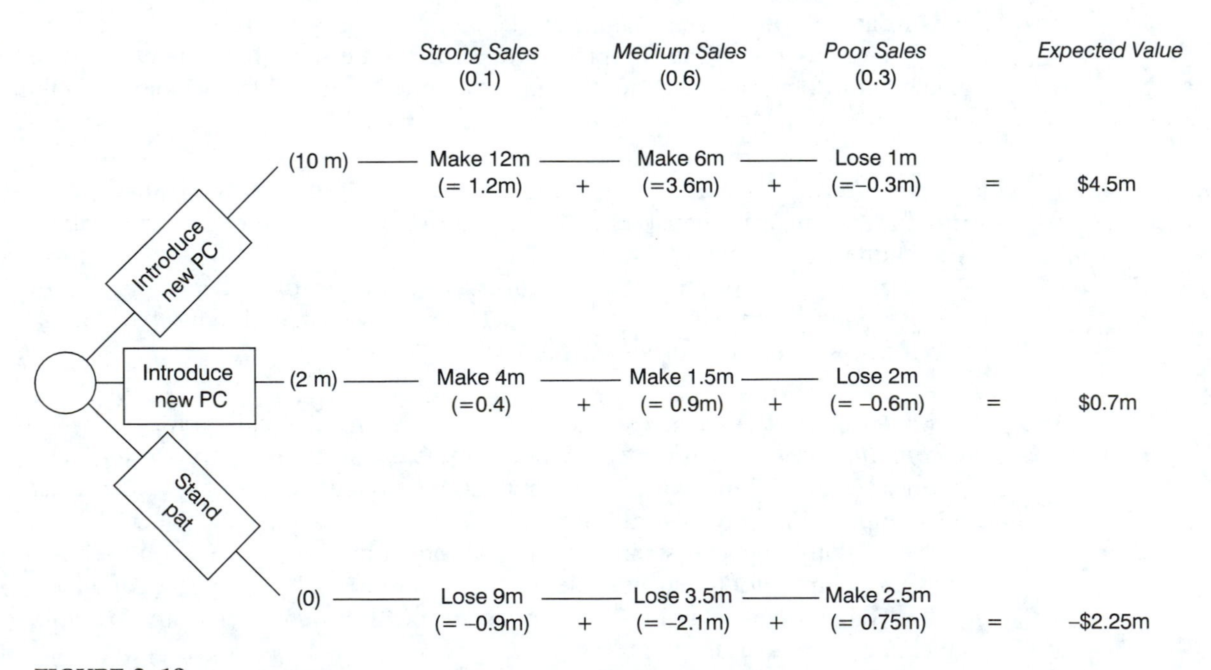

FIGURE 2–19
In evaluating alternatives, a decision matrix is one way to estimate the consequences of various courses of action—and inaction. (Note: All investment, sales, and expected revenue figures are in millions of dollars.)

Writing the Case Analysis

One of the unique elements of the written case analysis is that it requires no introduction or background. A reader of the analysis is presumed to have read the case first and to be familiar with its details. Your mental set as you begin to write should be to respond to a two-part question: "What's wrong here, and what should be done about it?" An effective response would be a slightly modified version of the analytical process you have already conducted.

An Outline for the Written Case Analysis

1. Define the problem or issue.
2. Present the solution.
3. Defend the solution.
   a. Outline other alternatives.
   b. Show why your choice is best: compare outcomes.
4. Present the implementation details.
   a. Who?
   b. When?
   c. Methods
5. State the expected results in a measurable way.

Notice that the written analysis is largely a reporting of the results from your mental analysis, except that you also include a defense of the proposed solution. In other respects, the organization is top-down and straightforward. (For a similar approach, see Unit 2.7, Proposals.)

Some final considerations: Though you may be writing the analysis for a management professor, the assumed reader is the person who evaluates your suggestions in the case situation (a senior decision maker). Your tone therefore should be serious, and the level of your language suitable to someone familiar with the company's affairs though not necessarily with the technical details of its products or services. Because you are writing to a busy executive, make your presentation clear and concise.

Example: A Case for Analysis

Environgard Inc. sells pollution control equipment to industrial firms to help them comply with federal and state environmental regulations. Its main product line is the PCS-200 series, a smokestack scrubber unit based on a system of chemical filters that remove sulfur from factory emissions. The PCS-200 is backed by a force of "consultants" who not only perform nominal service but also do extensive environmental counseling with customers and the general public.

Environgard also sells industrial safety equipment and provides consultation and services for the cleanup of hazardous wastes and for the redesign of production facilities to minimize pollution. It is a medium-sized company in its field, with sales of $149 million last year.

When Environgard was founded in 1962, it was the pioneer in a field that had not yet attracted notice or drawn legislation. The founder, C. Reed Griffith, was a chemical engineer who had been deeply affected as a youth by the 1948 incident in which some 20 deaths had been attributed to a period of noxious smog in Donora, Pennsylvania. Reed formed the company in large part because he wanted to help clean up the environment as well as make money.

The early years of Environgard were a struggle as Griffith oversaw development of new products and traveled extensively to build awareness of the deteriorating environment. He called on numerous companies and spoke to civic groups. His idealism and strong convictions won many admirers and even a few orders but left their strongest imprint on the company itself. Griffith's message to his associates stressed the following themes:

> We are not in this just to play on people's fears in order to make a quick buck. The environmental crisis is real, and it will get worse before it gets better. We're in this for the long term, and we mean to make a difference. And what we have to sell is not just a product but a means of reclaiming corporate responsibility. It's important that we offer quality products, consultation, and a service program second to none. It's even more important that we impress on our customers the stake we all have in reducing damage to the environment.

The company thus developed a strong sense of mission and, following Griffith's lead, an impatience with employees or customers who proposed cutting costs to deliver a cheaper product. Environgard products were well built, expensive, and supported by extensive service and consultation.

As environmental consciousness rose in the 1960s, the company experienced satisfactory growth, and its message of corporate responsibility seemed to find a better reception. After the Nixon administration formed the Environmental Protection Agency (EPA) in 1970 and Congress passed legislation limiting air and water pollution, Environgard's sales increased markedly.

The regulatory guidelines, however, also led to a sharp rise in competition. Environgard watched dozens of companies enter the field with products that undercut their prices. "These Johnnies-come-lately," said Griffith to his customers, "are quick-buck artists. They'll sell you a cheap unit, but when something goes wrong they won't be so easy to find. They'll sell you a filter box but not a strategy for long-term emissions control."

Some of Environgard's customers, however, switched to the competition. "Look," they said, "our costs of production keep going up. These new units are a lot cheaper than yours, and while they may not be as good, they get the job done. They're designed to get us in compliance with federal and state standards. We'd like to do even better, but the costs are killing us."

This philosophy of compliance with minimal standards left Griffith frustrated but did not reduce his determination. Environgard would continue to take the "high road" because they weren't in it just for the profits.

And indeed, Griffith's negative view of the competition seemed prophetic. Many of the new companies floundered, while Environgard maintained a steady course. Yet, some of the competitors did better. They expanded their market share and even improved their products—still designed for the compliance level but more efficient and reliable. The battle grew no easier. Environgard's bottom line in the early 1990s is shown in Figure 2–20.

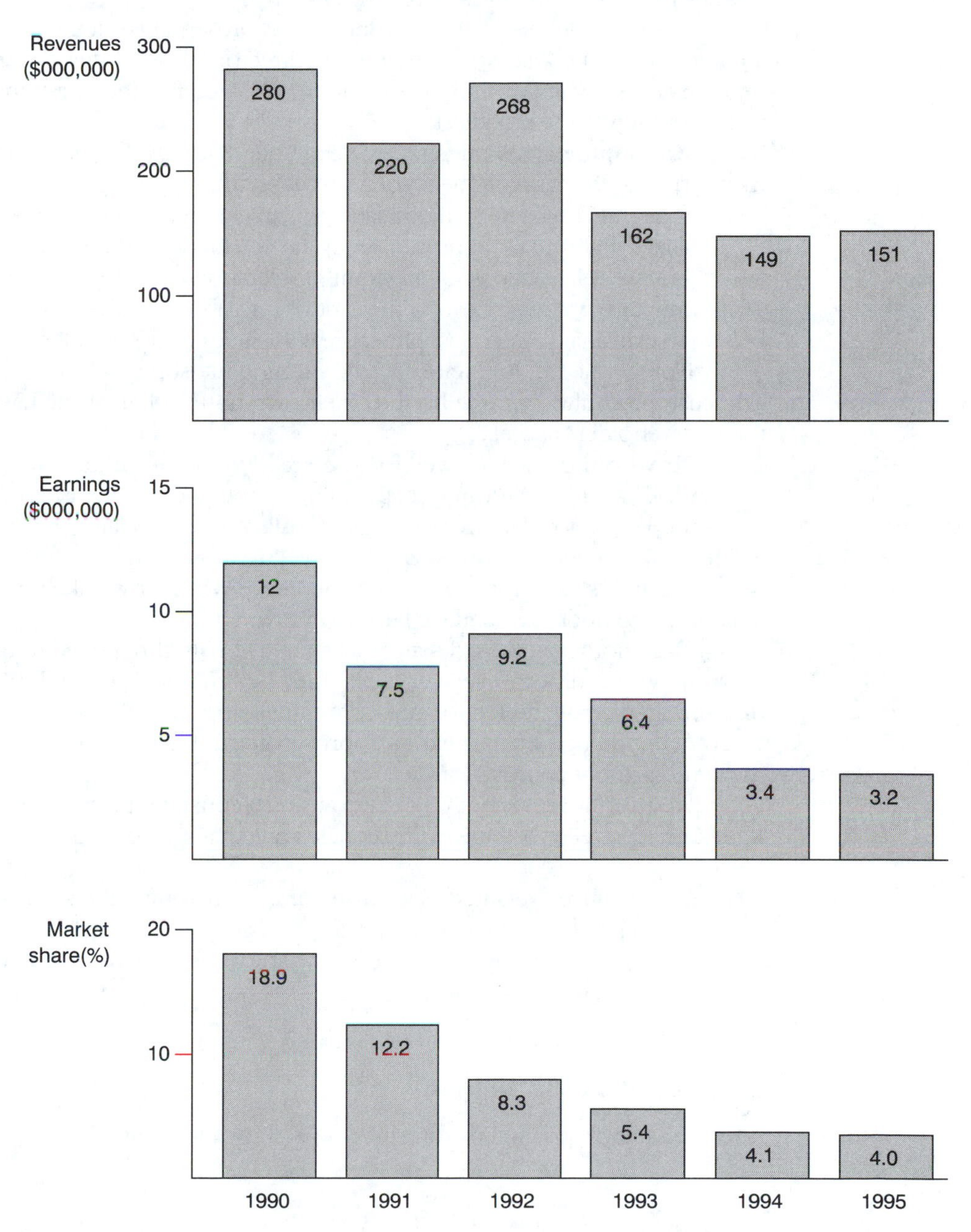

FIGURE 2–20
Revenues, earnings, and market share for Environgard, 1990 to 1995.

The steady erosion of sales, earnings, and market share took its toll on the employees. Some grumbled about the persistent cost-cutting measures adopted by the company. Environgard retained a core of steady customers, but their number was shrinking and new sales were few.

When one of her most promising sales representatives left the company to join a competitor, Kate Zaleski, the marketing director, decided that things had gone far enough and sought a meeting with Griffith. This was not their first encounter on the issues, but it was the bloodiest.

"We have to rethink our strategy, Reed," said Kate. "We're just not competitive anymore. The fact is that customers want a scrubber that will keep the EPA off their backs, and they can find half-a-dozen models in this market for up to 50% less than the PCS-200."

"That's the band-aid approach, Kate, and you know it. The EPA regulations are only a stop-gap. We sell solutions to long-term problems. Our service people and consultants offer companies value far beyond the cost of a scrubber. Maybe if our sales force did a better job explaining our philosophy, more companies would be willing to take us on."

"Not fair, Reed," said Kate. "We can argue philosophy till we're blue in the face. The customers always come back to their own bottom line. They have to answer to their shareholders as well."

"To whom do we answer for the steady decline in air and water quality?" demanded Griffith. "We can't keep aiming at the short term, quarter after quarter, while the long-term damage accumulates. All of us, especially our industrial clients, will pay a far bigger price someday unless we do more now."

"It's an 'us' and 'them' issue, isn't it, Reed? Whether we like it or not, *they* have a different point of view, and *we* have to live with it."

After another hour of debate, during which Kate threatened to quit and Griffith used the word "disloyal" several times, they held fire long enough to agree on conducting an overall review of the company's strategy. The review would decide whether to change Environgard's approach to its business or seek better ways to make the present approach succeed.

Reed Griffith reluctantly agreed to using a consultant to assist with the process. Kate Zaleski agreed to abide by the conclusions that would be drawn.

1. You have been retained as a consultant by Environgard to study the situation and to present recommendations.
2. Based on your analysis of the case, what would you suggest to Reed Griffith and his management team?

Example of a Case Analysis

A written analysis of the Environgard case is shown in Figure 2–21.

Problem

Solution

Defense of Solution

CASE ANALYSIS: ENVIRONGARD INC.

Environgard is a company with a civic conscience and a strong sense of its mission. It is also a business. The problem for Environgard is that it can't make up its mind which it wants more: to make money and grow, or to preserve the environment. It approaches the business of selling smokestack scrubbers by trying to awaken the conscience of its customers and get them to do more than they have to.

This confusion of purposes is most apparent in the company's pricing approach to its products. The PCS-200 sells for as much as double the price of a comparable unit, not because it is twice as good technically but because it is being used to subsidize the extensive service and consulting programs of the company. And even if the PCS-200 were twice as good as the competition's products, the latter are *good enough* to meet EPA specifications.

Four-Point Program

My recommendations to Environgard are as follows:

- Clarify your purpose as a company: Are you a business or an advocate for environmental preservation?
- Separate the various products and services you offer; allocate to each its true cost.
- Set appropriate prices for different products and services, and market these competitively.
- Use the company's image and reputation for environmental concern to support its marketing of products and services; don't use them to generate hidden costs to customers.

Analysis of Proposals

The first recommendation is really meant to emphasize a foregone conclusion. Even if Environgard dedicates itself to environmental protection, it must remain a viable business to finance that effort. The alternative would be to go out of business and reorganize as a nonprofit, public service agency. Such a step is likely to reduce the level of revenues available to carry out the organization's work.

As a business, then, Environgard might consider other ways of improving its financial position. It could seek to lower production costs as a prelude to lowering product prices.

Alternatively, it could seek to find more customers as a means to increasing unit sales so that unit costs could be lowered.

FIGURE 2–21
Case analysis (pp. 133–134).

Both of these alternatives seem unlikely, given the realities of a much more competitive industry since the 1970s. Although some competitors have left, others have settled in for the long haul and are offering reasonable quality and service for a lower price. Environgard is unlikely to pull any production miracles or breakthrough technologies out of its hat.

As another alternative, some members of the company have suggested that Environgard's marketing force should find more effective ways to "sell the company's environmental philosophy." Although this may be possible, it would run counter to the strategy of setting true costs and staying competitive in the pollution control industry on a business basis.

The recommendations presented here appear to be the company's best options for restoring its financial health and continuing to make a significant contribution to environmental protection.

Implementation

Working the Plan

To achieve this strategic change, Environgard's president should take the lead by explaining the new direction to his department managers and asking them to develop appropriate plans for their functional ares. The new strategies should then be presented to all employees in a series of one-day assemblies at which questions and answers would be freely exchanged. The assemblies would then break into smaller groups headed by the departmental managers to discuss implications for their areas.

It is estimated that this process might take about 3 months to complete because of the need to finalize details as well as to continue present business activities. During this period, special emphasis should be given to the development of a marketing program that would present the company to its customers in a new way. If possible, preliminary versions of advertising and sales brochures should be introduced at the employee meetings to solicit feedback. It is vital that employees be consulted and listened to throughout the process.

The necessary design and production changes, although not likely to be extensive, would require about a year. The start of the new marketing campaign should precede this date by about 3 months.

Expected Results

Conclusion

If these recommendations are adopted, the company should realize a 25% growth in revenues and a 20% increase in market share in the first year of the new program. Earnings growth that year is likely to be modest (in the range of 2% to 4%) because of the costs associated with the change. In the 3 years after that, however, earnings may be expected to grow to at least $10 million. This is a conservative estimate based on the current ratio between earnings and sales.

FIGURE 2–21, *continued*

Unit 2.9 Case

Established in 1955 as a neighborhood bookstore, Fit to Print Books Inc. has been in the Rose family for three generations. Each of its proprietors has reflected his own interests and those of his times in guiding the enterprise over its 40-year existence.

Fit to Print Books was the brainchild of Lester Rose and his partner, George F. Fisher, a friend at the University of Chicago in the early 1950s, where they decided, in the face of their frustrations with the campus bookstore, that a store offering quality selections at reasonable prices could be an interesting and profitable venture. Lester, an avid reader and participant in campus literary and cultural affairs, felt he could bring interesting new writers to a sophisticated college audience. George, a business major, brought a financial and marketing perspective to the partnership.

The fledgling business did well in the 1950s on the strength of solid profit margins maintained on books that were perceived by the public as the work of serious writers with avant garde ideas. The Friday night poetry readings organized by Lester were lively and well attended. Books cost a little more at Fit to Print, but the selection and quality were worth the markup.

By the 1970s, George's half of the business had been bought out by Lester, as George got a divorce and moved to Laguna Beach, California. Lester, too, had moved and taken the bookstore to Oak Park, a Chicago suburb (and birthplace of Ernest Hemingway). In 1973, Lester brought his son, Willard, into the business as a junior partner. Willard was a recent graduate of the University of Wisconsin at Madison, where he had participated in protests against the Vietnam War. After graduation, he had lived in a commune in New Mexico for several years.

On his return, Willard gradually took over the business from Lester and shifted the product line into the areas of youth culture and Eastern philosophy and religion. He also introduced a music department featuring 1960s rock groups, and added paraphernalia such as posters, incense sticks, bead curtains, and Flower Power bumper stickers.

The store did reasonably well in the 1970s and 1980s as Lester retired and Willard assumed full management. In 1985, Willard moved the store to a shopping mall in Schoenberg, a new subdivision beyond O'Hare Airport. In the early 1990s, a third-generation Rose was brought into the business, in the person of Wendy, daughter of Willard by his first wife. A 1989 marketing graduate of Concordia College, Wendy had been groomed to assume control within a few years, but by this time, Fit to Print had become an unsteady enterprise.

Independent booksellers had come under pressure from high-volume chains such as Borders, Barnes & Noble, Super Crown, and others. After the fading of the youth culture of the 1960s and 70s, Fit to Print had relied on the formula of offering something for everyone—bestsellers, romance novels, cookbooks, thrillers, science fiction, how-to manuals, and dozens of other categories. Many of these were items that Willard hadn't read himself and didn't care about. His strategy was to cover all the market "segments" and count on steady customer traffic to keep his cash registers humming.

This strategy seemed increasingly in need of review. In the past five years, financial results had been steadily eroding, as shown in the following graphs:

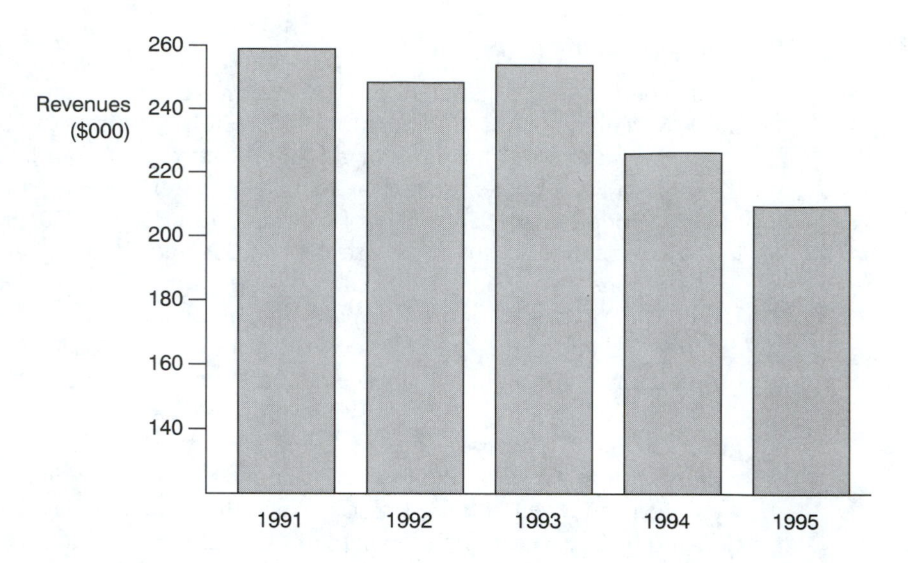

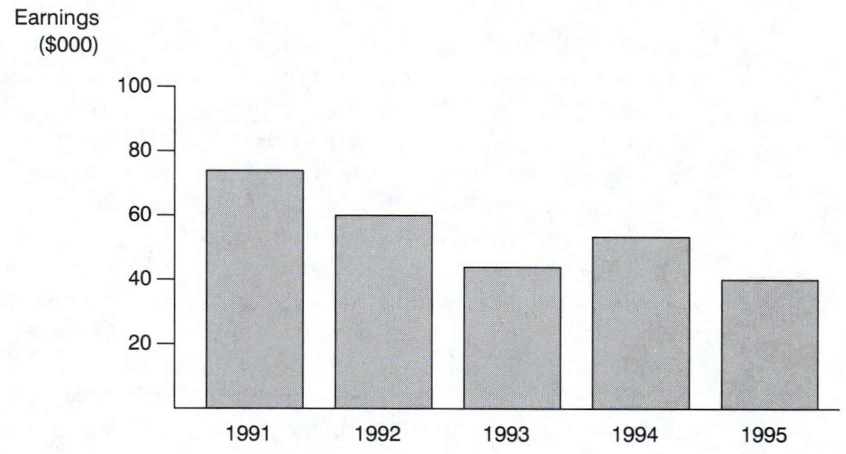

Within this generally bleak picture, however, a few bright spots were visible. Willard's interest in computers was reflected in an excellent and broad selection of books and manuals from this field. Sales of computer items, including software, had grown every year during the general decline, accounting for the following proportions of revenues and earnings in the first half of the 1990s:

	1991	**1992**	**1993**	**1994**	**1995**
Revenues	25%	32%	38%	40%	42%
Earnings	30	35	40	45	49

"I don't think we can compete with the Barnes & Nobles as a high-volume, comprehensive bookstore," said Wendy to her father, "but we might survive if we were willing to change. We might have a future as a computer specialty store—books, manuals, software, *and* hardware. You have the interest and expertise in this field, and I think I could market such an enterprise successfully."

"I don't know, Wendy," said Willard. "I don't know if I can let go of a family legacy, which is *books,* 'fit-to-print books,' as my father used to say. A bookstore is something special—a world of ideas, a storehouse of human creativity and imagination."

"I understand, Dad," said Wendy, "but for some time now we haven't really embodied what either your father started, or what you tried to do in the 1970s. We've tried to be all things to all people without making sure that everything we did was first-rate. It's time for a change."

"I don't know, Wendy. What would we call ourselves? We couldn't be a 'bookstore' anymore."

"Let me write it out as a case analysis with a proposed solution, Dad," said Wendy. "I did a lot of those in college, and they help sort things out. Then you can decide if it makes sense."

If you were in Wendy's shoes, how would you write the analysis, and what would you propose to Willard Rose?

SECTION 3
Correspondence

Letters and memos—the main subjects of this section—are the most frequent kinds of business writing by far. They are a necessary link within and between organizations of all kinds. They are likely to be used even if reports or other summaries of information are required:

- ☐ A cover memo may present a feasibility study to senior management.
- ☐ A cover memo may send a requested schematic diagram to a co-worker.
- ☐ A letter of transmittal may convey a formal proposal outside the organization.
- ☐ A biweekly progress report to a supervisor may be sent on a memo form, with copies to project-team members.

Business letters are also the common means by which private individuals interact with organizations—in requesting information, ordering products, registering complaints, or even seeking entry to the organization via a job application. And as private individuals, all of us are familiar with the sales and promotional letters that businesses or nonprofit organizations send us.

In the electronic realm, the familiar types of correspondence have undergone something of an identity crisis. Is an e-mail message a "memo" or a "letter"? The message is set up with *To-From-Subject-Date* headers like a memo, but it is typically "signed" at the bottom and sent outside the organization, like a letter. We can say, at least, that e-mail is correspondence—the sending and receiving of messages—rather than a form of reporting. E-mail will be included here, therefore, because it has features of both letters and memos.

Letters and memos are worth studying because they are basic to doing business, getting work done, and even to getting into business. They are also a specialized form of communication. You could say they are deceptively simple.

Good memos are often brief and informal. Good business letters are direct, focused, and written in plain language. Neither is so by accident. Efficiency and plainness suit their purposes and get the job done. Busy readers want messages to come to the point immediately and to include all necessary information but no more. And readers appreciate courtesy and the relaxed tone of ordinary conversation.

What is deceptive about good letters and memos is that writing them takes as much care and concentration as writing complex reports. Conciseness, completeness, and efficiency are never easy to achieve. Like good reports, good letters and memos require an understanding of purposes, audiences, and strategies. And they require lots of practice. The units in this section will give you the understanding that makes practice meaningful.

UNIT 3.1 MEMOS

Writing memos provides experience with one of the most common and useful forms of business communication. It also trains you to think in a top-down way about a subject to get your point across in the most efficient way.

POSSIBLE ASSIGNMENTS

Send a memo to one of the following:

- ❑ Your instructor on the status of a class or lab project
- ❑ Your instructor with questions about grading, course materials, or class policies
- ❑ Your classmates to coordinate a group project
- ❑ Other students to announce a meeting or handle other club business
- ❑ The library, an administrative office, or an academic dean to request information or suggest improvements
- ❑ Coworkers with conflicting schedules to arrange a meeting
- ❑ A building or maintenance supervisor, pointing out a needed repair or unsafe condition

Approximate length: A few lines to a few paragraphs.

Explanation of a Memo

The memo is an informal and efficient way to send messages within an organization. It provides a record for all concerned. Because sender and receiver know each other and know the organization, there is no need for introduction or background explanation. You can get right to the point of the message.

The first sentence of the memo should state your main point or question. Following sentences then provide any supporting information or clarification for the first sentence. This is *top-down* organization, and it suits a busy reader. You are more likely to be heard as a result.

To write top-down, you will often have to retrace your mental steps or reverse the order of events as they happened:

Day 3.	The coffee station runs out of coffee bags by midmorning, for the third day in a row.
Day 5.	The vendor has stocked the coffee station with a peculiar-tasting off-brand.
Day 8.	The coffee machine is having an oil spill; there are globules of grease in your cup.

Your main point here is a conclusion you have reached about the situation. Your memo to the Human Resources Department states your case this way:

> We need to replace the coffee service with a more responsible vendor. The machine is fouled up and dispensing oily coffee. When it does work, we usually do not have enough coffee bags to go around, unless they are Brand X, which nobody drinks.

Memos can run to many pages but are commonly no longer than three to five sentences and sometimes consist of only one. (For example: "The office will be closed Monday, September 6, in observance of Labor Day.")

The message is usually typed on a preprinted form. These forms can vary in size and design but typically bear a company logo and the printed words *to, from, date,* and *subject* at the top. The memo form of a company called ABC appears in Figure 3–1.

While the memo format in Figure 3–1 is serviceable, a number of variants are common, including those shown in Figure 3–2, where a vertical arrangement of elements is used.

When preprinted memo forms include only the logo, the writer must decide how to arrange the *to, from, subject,* and *date* entries. In that case, the format shown in Figure 3–1 offers several advantages over the vertical approaches of Figure 3–2.

Notice that in Figure 3–1 the *to* heading provides extra space for names. This lets you address more than one person in the memo. One of the advantages of this format is that groups with common purposes, such as departments or committees, may be easily covered by a single message.

You can also write a memo to one person and *copy* several others. Typical "others" might be your supervisor or co-workers, or those of the person you address. Use a "c:" designation for copies. Here again the horizontal layout in Figure 3–1 better accommodates the purpose than the vertical mode. In the latter case, copy information must be added at the end of the memo, where it may receive diminished attention.

ABCo

Blenheim Tower
4 Wellington Drive
Waterloo, Ohio 43208 614/568-1200

Interoffice Memo

TO: **FROM:**

SUBJECT: **DATE:**

FIGURE 3–1
The ABC interoffice memo form.

Unless your message applies equally to a group, write your memo to one person and copy the others. Otherwise, you may divide responsibility for action and lessen the chance of response from anyone.

Before sending your memo, take a second look at the *subject* entry you have devised. A well-designed statement of your subject and purpose can gain attention for the memo and help the reader into the message. Design your subject statement with the following aims in mind:

- ☐ Name your subject and say something about it that reflects your purpose.
- ☐ Try to state the whole point of your memo in a single phrase.
- ☐ Construct the subject phrase as you would a subhead in a report: Use nouns and modifiers, but no verbs.

For an illustration of these principles, consider how you would design the subject statement of your memo about the coffee service on page 141. Which of the following statements would you choose?

Subject: Irresponsibility
Subject: Coffee Service
Subject: Poor Coffee Service
Subject: Replacing the Coffee Service

The first statement may gain attention, but is too vague to help the reader into the memo. The second statement names the subject, but then says nothing about it. The third version comes closer to your purpose, but does not actually say what you want done. The final, and best, version summarizes the memo by naming the subject and indicating your purpose. (Note that the final phrase uses the noun *replacing,* and does not rely on a verb to express the purpose.)

Review your memo from the standpoint of style and tone as well. To help you choose the right level of language, think of all the people you listed after the *to*

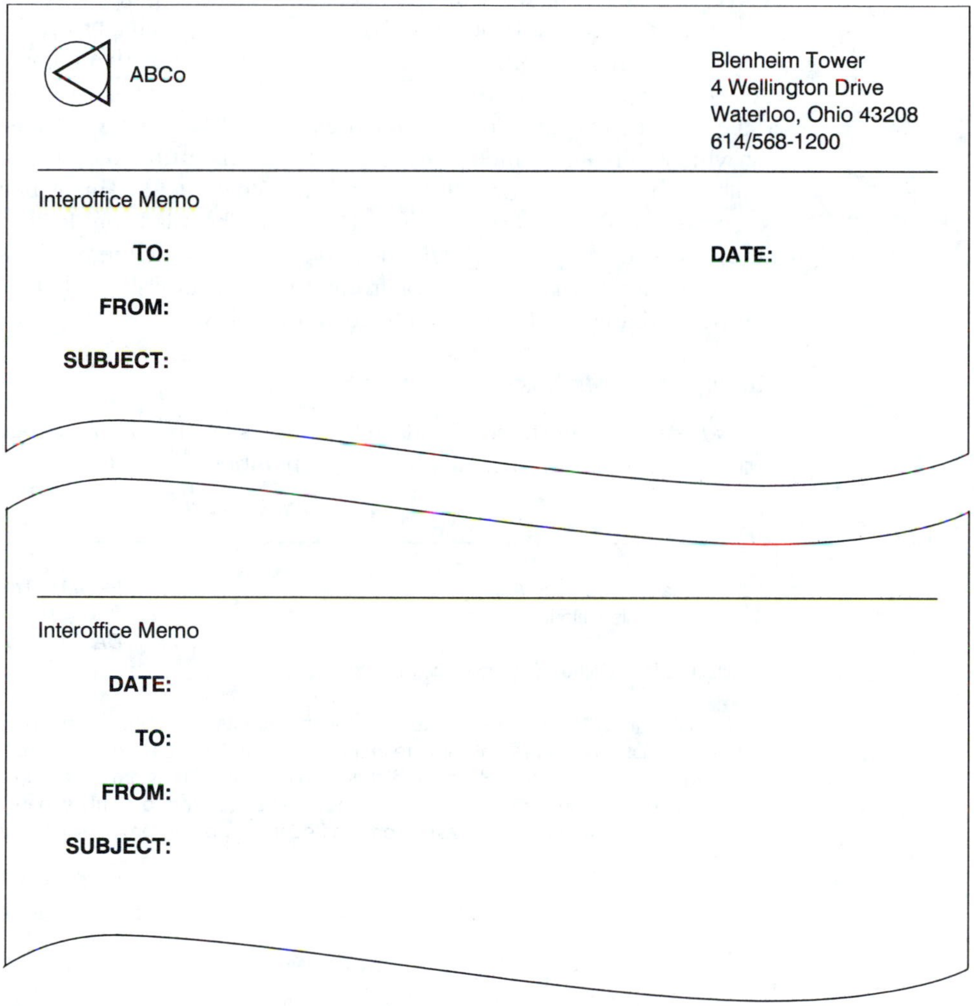

ABCo

Blenheim Tower
4 Wellington Drive
Waterloo, Ohio 43208
614/568-1200

Interoffice Memo

TO: **DATE:**

FROM:

SUBJECT:

Interoffice Memo

DATE:

TO:

FROM:

SUBJECT:

FIGURE 3–2
Two common variants of the basic memo format, featuring a vertical array of the orienting elements.

heading, all those you copied, and yet others who might read the memo because it was passed on to them. Because memos are an open form of communication and are widely shared, it is best to keep the language nontechnical, the sentences and paragraphs short, the message to the point. Highlighting techniques such as bullet points and charts or tables can effectively replace straight narrative.

Similarly, the tone of your message should be professional and cool, even when you are upset or feeling pressure. When memos are about problems, they are often passed on to others for comment or approval. If your memo accuses, rather than explains, or insists, rather than suggests, your purpose may be frustrated by readers who conclude that you are unfairly attacking or blaming others for your mistakes.

Consider the case of Trudy Bantra, a training manager for Blaustein Systems. Trudy gets a memo from the Accounting Department, with a copy to her boss, Louis Delorio. The memo questions an expense claim for a rental car filed by Jerry Dillbert, a sales trainee in one of Trudy's training groups. The claim contradicts company policy, which calls for trainees to use courtesy vans and taxis. The memo concludes: "Dillbert says you authorized the rental car. What gives?" This is news to Trudy, who is tempted to fire off a reply saying that Mr. Dillbert is lying. Instead, she composes herself before composing the memo and then sends the memo shown in Figure 3–3. Would you agree that the tone of Trudy's memo establishes a better response to the complaint than would pointing a finger at others? Why?

Examples of Memos

Class Projects. Professor Westlake has requested that all project groups in his class report on the status of their projects. A member of one team, Charles King, sends

TO: Clara Schumann/Acctg.
c: Louis Delorio

FROM: Trudy Bantra/Trng.

DATE: Nov. 2, 1997

SUBJECT: J. Dillbert Expense Claim

Mr. Dillbert may have misinterpreted the materials sent to him prior to the training session. The materials say that rental cars are authorized for outside salespersons, not for trainees during the training phase. Because of the large number of new employees in the session, I was not able to speak personally to each person beforehand to clarify expense policies. I regret the error and I hope it hasn't complicated the reporting process too much.

FIGURE 3–3
Rental car expense claim memo.

TO: Prof. Westlake

FROM: Charlie King

c: Ruth Bentkowski
Ellis Jackson
Anne Perlman
Doug Sorvino

DATE: April 8, 1997

SUBJECT: SY-200A Project Status—Group 5

Our project is now in the programming stage. On Tuesday, we divided the program into five sections for the members to work on individually. We plan to complete this stage by April 18 and meet April 19–21 to merge the results. If team members complete their assignments, we will be on schedule for the final submission of the report by April 30.

FIGURE 3–4
Class projects memo.

the memo in Figure 3–4 and copies it for his classmates. (Notice how the last sentence indirectly reminds team members of their responsibilities.)

Library Improvements. Elise Fuhr, a second-year electrical engineering student, is frustrated by her inability to complete lab projects on time. Twice this term, essential reference materials in the engineering library have been missing or checked out to other students or faculty. The faculty are even worse than students in keeping materials beyond the due date. Elise feels she is being penalized for others' actions and for unwise library circulation policies. She composes an angry message to the librarian, Ivars Stumbris, complaining about the loan policy and the lack of a security system. A roommate reviewing the draft tells her it needs revision: There are too many different topics and the tone is too emotional. Elise accepts the advice. She revises her draft and types it out in the form shown in Figure 3–5.

Unit 3.1 Exercise

1. Professor Westlake has asked you to write a memo to Joyce Brotherly, director of the Career Counseling office, asking her help in finding a guest speaker for the EE-384 class to talk about careers in electrical engineering. He feels that a recent alumnus might be a good choice for the task, and suggests that you copy the memo to Hugh Reynolds, coordinator of the Alumni Association. Both Joyce and Hugh report to the Dean of Students, Nick Copernicus. Select an appropriate memo format and design a message to meet Prof. Westlake's request.

MEMO

DATE: April 3, 1997
TO: Dr. Ivars Stumbris, Library Director
FROM: Elise Fuhr, EE student
SUBJECT: Restricting Circulation of High-Use Materials

The lab projects in Prof. Westlake's Digital Systems course (EE-384) would be easier to complete if several sets of reference materials were placed on three-day reserve behind the front desk. Open circulation of these materials, as well as lack of a library security system, create a high probability that essential manuals and manufacturers' guides will be checked out or even missing.

My classmates and I would appreciate restricted circulation for the following materials:

- National Semiconductor *CMOS Logic Databook* (1995)
- Texas Instruments *MOS Memory Data Books* (1988–1995)
- Texas Instruments *The TTL Data Book*, Vols. 1–4
- Advanced Micro Devices *PAL Data Book* (1992–1996)
- Motorola (McGraw-Hill) *Microprocessor Applications Manual* (1994)
- Hayes, *Digital System Design and Microprocessors*, 2nd ed., McGraw-Hill, 1992

c: Prof. Thomas Westlake

FIGURE 3–5
Library improvements memo.

2. You bought a parking permit at the start of the semester for the West End lot, the only site within walking distance of the Engineering building. You need quick access and exit to manage both classes and a part-time job. Several mornings, no parking spaces are left when you arrive, and you are forced to search for space in the street. Other mornings, some cars are straddling two spaces and all other spots are taken. You've been late to Prof. Masterson's Transform Analysis class four times already, and she has taken to calling you The Late Mr/s. __________. You complain to the watchman in the campus security office, who tells you he can't do anything. He has to monitor calls at his desk, and can't patrol parking lots around campus. "Take it up with my boss," he says, and gives you a business card for Dana A. Trevino, head, campus security, who apparently reports to Clark Heisler, the vice president of administration.

Design a memo that will have the best chance of achieving the dual aim of (a) guaranteeing your rightful access to the West End lot and (b) improving your standing with Professor Masterson. Send it to the person or persons who can best serve your purposes, with copies to others you wish to read your message, if any. Be prepared to justify your strategies.

3. Assume you have changed your mind about the parking problem in Exercise 2. You decide to press your campaign through the Student Government Association to create pressure on the administration to improve parking availability near campus and to better enforce parking regulations. How would your message change, if at all, compared to your approach in Exercise 2? Whom would you address in the student government, and whom would you copy, if anyone? Discuss.
4. Which of the following subject statements is best for a memo about your parking problems addressed to Dana A. Trevino?
   - **a.** Betrayal of Trust
   - **b.** You Need to Live Up to Contractual Agreements
   - **c.** Assuring Spaces in the West End Lot
   - **d.** Lack of Spaces in the West End Parking Lot
5. Would the best subject statement from Question 4 also be suitable if your memo were adressed to Vice President Heisler? To the student government president? Explain.
6. Look at the office service memo you wrote the Human Resources Department (p. 141). Suppose the coffee wasn't provided by an outside vendor but by a staff member in Human Resources, whose duties included stocking coffee bags, filters, and related items and providing service for the machine. Would your memo change in these circumstances?
   - **a.** Would you send it to someone else in the organization?
   - **b.** Would you change the language of the memo? Is the suggestion that the service person is not "responsible" too inflammatory, for example? Discuss and illustrate.

UNIT 3.2 ELECTRONIC MAIL

This unit looks in a practical way at a basic service of the Internet—the provision for electronic mail—and offers suggestions for how to get the most out of this service in sending and receiving both individual mail and postings to electronic seminars or discussion groups.

The Internet Phenomenon

The increasing importance of the Internet can be suggested by a few statistics pointing to its phenomenal rate of growth. While the number of people actually using the Internet can only be estimated, one set of figures points to 38 million users in 1994, 56 million by the end of 1995, and a projected 200 million by 1999. The number of host computers (primary machines to which many other computers are connected) rose from 1.8 million in July 1993, to 3.2 million in July 1994, to 6.6 million in July 1995. By the end of the 1990s, 120 million host computers are expected to be connected to the Net (see Figure 3–6).[1]

1. Cynthia Bournellis, "Internet '95: The Internet's Phenomenal Growth Is Mirrored in Startling Statistics," *Internet World*. (November 1995): 47.

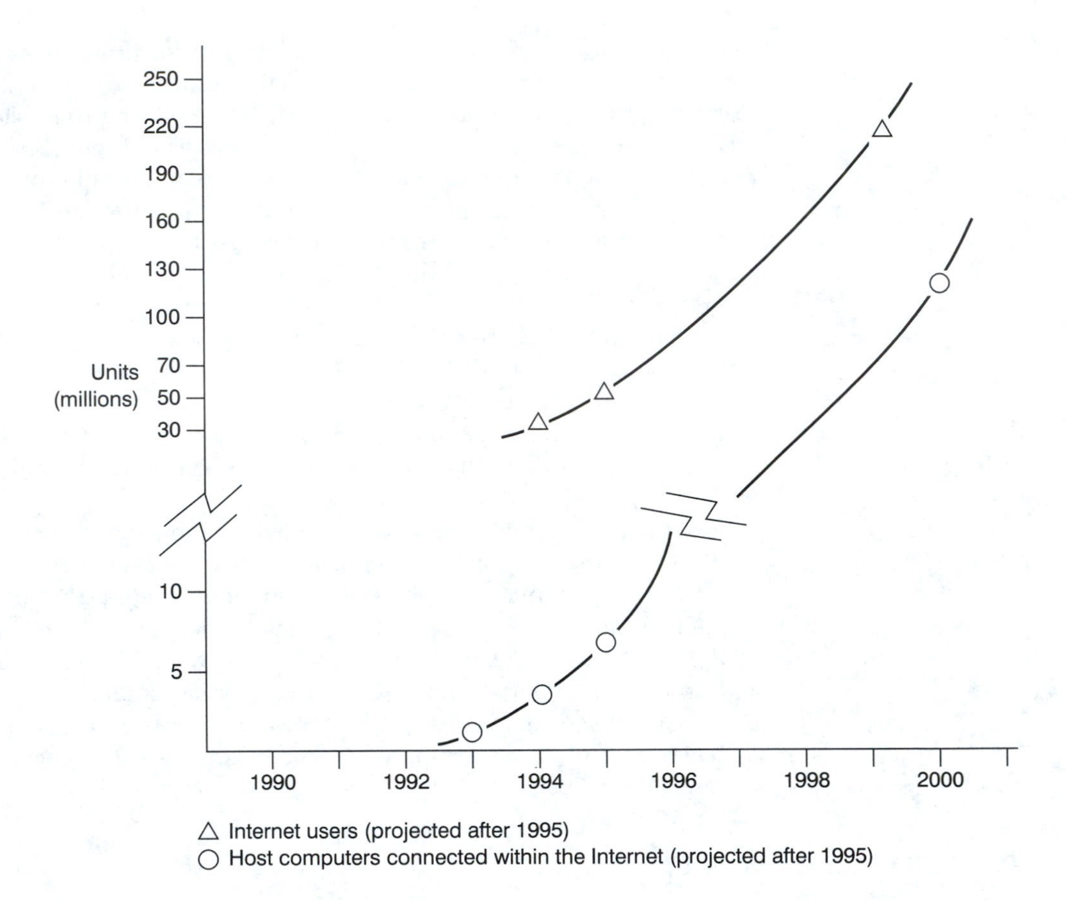

FIGURE 3–6
Exponential growth of Internet users and host computers.

This kind of growth suggests that people worldwide are finding the Internet an interesting and *useful* medium. With respect to a widely used service of the Internet, electronic mail, it is not hard to understand why this is so.

Consider the memos discussed in the previous unit. Having dealt with the issues of how to format and compose a variety of messages, the memo writer must then put these thoughts on paper, which must then be sent to the recipient. Sending a message within an organization is a time- and labor-intensive process. Someone must take the memo from the sender to the receiver, usually with intermediate steps involving a central clearing mechanism such as a mail room and its staff.

The same memo entered into computer memory and sent electronically is swiftly routed to a recipient's e-mail box, whether it be in the next office, on the next floor, or on the other side of the planet! And, as Internet users are fond of pointing out, the e-mail memo hasn't "killed any trees" that are used to make paper. Using the electronic medium does change a few of the word-processing capabilities, and requires that the user observe some guidelines that can make e-mail a practical and receiver-friendly tool.

Basic E-Mail

Your connection to the Internet may be provided by your college or university, by your employer in a commercial account, or by a private subscription service such as Prodigy, CompuServe, or America Online. To get on the Internet, you access the organizational network that is actually connected to the Net or use the modem serving your computer to call up a host computer that is actually connected to the Net. You log in, using a prearranged address and password that are handled by your e-mail program. The actual sequence of steps may be similar to the steps given in Figure 1–10 on page 54–55.

After you have logged in, your e-mail program allows you to send, receive, and reply to messages, along with some additional services, such as handling printing instructions and filing incoming and outgoing messages into folders that the program helps you create.

To help you *send* an e-mail message, the program presents a template prompting you to key in header information, equivalent to the preprinted items on a memo form. As an example, the e-mail program Eudora presents you with a formatted screen, with about six or more lines at the top, as shown in the following model, in which the elements provided by the program are in bold and the elements keyed in by the sender are in lighter type.

HEADER

To: nhawkin@primenet.com
FROM: jhobbes@calvinaccess.com (James Hobbes)
Subject: Program Guide
Cc: nancyab@tng.njstevens.com
Bcc:
X-Attachments: T110PROG.GDE.DOC

\- -

MESSAGE

Hi, Natalie:

We were able to finish the T-110 Program Guide in draft form last week, but I think it needs serious editing and revision. I'm enclosing a copy of the guide as a Word 6.0 document. Could you take a look at it and give me and Nancy Abrams, my colleague from N. J. Stevens, some feedback, particularly on the installation and testing sections? We will owe you big time for any help you can give us.

Cheers,
Jim H

****

TRAILER

James M. Hobbes
Training and Development Specialist
Calibration Systems Inc.
486 W. Sumpter Ave.
Kingston, OK 52138

Phone: 819-834-4572 x2345
Fax: 819-834-3356

"Training is only the first step in development."
—C. Eduardo Domingo

In this example, James Hobbes e-mails Natalie Hawkins at her Internet address and copies the message to Nancy Abrams. James "attaches," or encodes as a text file, a copy of the program guide, which Natalie can print out as a Word 6.0 document by using her own word processing program. James's identifying address (and name) are provided by the e-mail program. James also sent a **blind** copy of the memo to his boss; this copy is "blind" because the e-mail program keeps this information from the recipient. At the end of the message, the program also prints a standard "trailer" of information that James can use as an electronic business card for networking purposes.

To **receive** this message, Natalie logged in and was told by her e-mail program that she had "new mail"; she acknowledged this information and opened the "In" folder of her mailbox, finding a list of messages received. One of these was from "jhobbes" on the subject "Program Guide." Natalie selected and then opened James's message, which was headed by information in the following format.

Return-Path: jhobbes@calvinaccess.com
Date: Fri, 20 Oct 1996 14:50:33 -0700 (MST)
X-Sender: jhobbes@calvinaccess.com
To: nhawkin@primenet.com
From: James Hobbes <jhobbes@calvinaccess.com>
Subject: Program Guide

In addition to repeating information keyed in by the sender, the e-mail program has added the precise time the message was sent, including its time zone, and has set up a return path (i.e., the sender's address) should the receiver wish to reply to this message. In that case, Natalie would only need to start the "Reply" utility in her system to be presented with her own reformatted message screen showing herself as sender and James as receiver.

Along with composing her own reply, Natalie could then delete parts or all of James's message to save memory and focus her response. She could delete lines by highlighting them and then pressing the Delete key. Using the "cut and paste" functions available in her system, she could also move parts of James's message around to allow more specific replies to lines or sections of the message. More specific identification of words or phrases can also be achieved by using a string of carets (^ ^ ^ ^) or tilde (~~~) signs to "underline parts" of the message to which she might want to reply.

Many normal word processing capabilities are not available in e-mail, however. There are no italic, bold, or adjustable fonts, and the emphases or inflections signaled by these features must be suggested in other ways: Asterisks may be used to *highlight* terms, partial _underlines_ to indicate italics, and UPPERCASE letters may suggest **bold type,** although capital letters also indicate strong emotion and should be used sparingly to avoid the impression that one is shouting.

The word wrap feature that automatically carries your words to the next line in word processing is not reliable on the Internet because of varying standards among the networks. It is good practice, therefore, to hit the return/Enter key at about 72–75 characters of line length yourself, so you don't get one- or two-word continuations all by themselves
on following lines, like the one above, that end abruptly and waste space.

Improvising for the lack of expressive capabilities, Internet users have developed a lexicon of symbols, including the following:

:-)	the basic Smiley (look at it sideways) to connote good cheer
:-(	the opposite of a Smiley
:-O	shock or surprise
;-)	a wink

Artistic users have also produced effects such as the following:

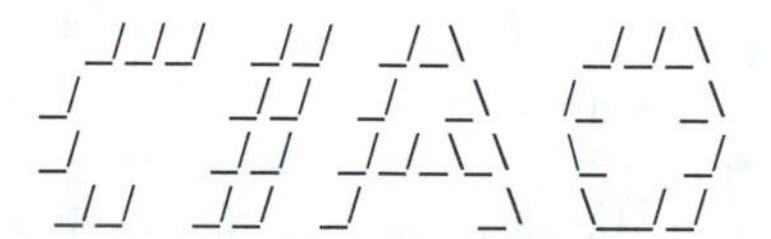

For those with the time and the inclination to generate such effects, the result is a pleasant enhancement of the standard Internet message.

Electronic Discussion Groups

Discussion groups on the Internet are also known as electronic seminars, forums, or conferences, or as newsgroups. Newsgroups allow participants to post messages that are distributed to all subscribers, and to receive or send comments, questions, interjections, and brief or lengthy rebuttals. Groups often combine beginners and experts, and the mix turns out to be productive, perhaps because it allows beginners to learn and experts to expound their expertise.

To join a newsgroup, send a "subscribe" message to the administrator (actually a computer program), who will have a different address than the one used to post messages to the group. To join the group, you would typically state: SUBSCRIBE [YOUR FIRST NAME] [YOUR LAST NAME]; to leave the group, or suspend participation, the typical request would be: UNSUBSCRIBE [YOUR FIRST NAME] [YOUR LAST NAME].

Upon joining a newsgroup, it's a good idea to create a folder and to save in that folder the instructions you receive about how to join and leave the group and how to send messages.

The themes or topics of discussion can be quite focused or very broad. Broad or specific categories in areas as science, politics (soc = social issues), business (clari.biz.top = top business news), computers, music, art, or sports are reflected in the following newsgroup names:

Broad	***Focused***
soc.politics	rec.motorcycles.harley (Harley-Davidson bikes)
soc.history	alt.chinchilla (on chinchilla farming)
soc.women	alt.skateboard

soc.travel	chi.eats (restaurants in the Chicago area)
clari.biz.top	comp.sys.mac.hardware (Macintosh hardware)
rec.music.classical	nj.weather (weather in New Jersey)

Academic departments and project groups in companies may create newsgroups focused on common tasks or themes. Such groups may use a local area network or the Internet itself to keep everyone informed and to generate new ideas.

Many newsgroups allow messages to be posted freely, but some groups are moderated, in which case submissions are screened or edited to minimize irrelevant, wasteful, or disruptive comments. If the group is not moderated, you can set up a "filter" mechanism that will automatically pull messages on topics of no interest to you and place them into a separate folder that you can scan later, for likely quick disposal of most items. Check the "Help" entry on the menu bar for instructions on how to set up filters in your e-mail program.

Among the most notorious of the disruptive comments are *flames,* which are inflammatory statements by members of discussion groups, and which may lead to "flame wars" that derail productive discussion into the arena of petty insults and recriminations.

To minimize the frustration that may lead to flame wars and other disruptions, subscribers should practice newsgroup etiquette, or *netiquette.* Ordinary thoughtfulness and politeness will go a long way toward this goal, along with a few practical suggestions:

1. Don't post a message that only attaches an "I agree" or other brief comment to someone else's posting. The result will be a repetition of not only the original message but all the introductory "header" information, forcing subscribers to spend time retracing their steps.
2. Instead, delete headers, along with unused sections of the prior material, and reference the previous message by identifying its author ("James Jones says that:") or by highlighting particular lines or phrases in the message for comment.
3. Consider whether you might not prefer to post a private response to some newsgroup messages, since the sender's e-mail address is included in the header. If you learn, for example, that the sender is a fellow graduate of your alma mater and you feel like chatting about favorite campus hangouts, do it in a private message.
4. Before asking a question of the group, check the Frequently Asked Questions (FAQ) file, and review the summaries of past discussion topics if these are available. This precaution will save time for other members.
5. Keep your sense of humor and don't be quick to take offense at comments by other subscribers. Use the Smiley symbol, :-), to suggest your comments should be taken in a lighter vein.

UNIT 3.3 LETTERS

In spite of advances in electronic communications, letters remain a basic tool for conducting business and are likely to be with us for some time yet. Between companies, letters serve the same role as memos do within the company; like memos, they offer several key advantages. A letter usually costs less than a long-distance phone call. It can be passed on to the person best able to help you. And it provides a permanent record of the transaction for you and the company. Your ability to write effective business letters will be a real asset to your career.

POSSIBLE ASSIGNMENTS

Write to one of the following:

- ❑ Businesses for product specification sheets, catalogs, schematics, software information, annual reports, or similar information
- ❑ Companies or schools to determine what systems or equipment they use for particular purposes
- ❑ Organizations to get information on trade shows, seminars, conferences, conventions, or meetings
- ❑ Professional societies about student chapter activities and programs
- ❑ Manufacturers with your questions or comments about equipment you have used in school projects
- ❑ Individuals or companies to engage guest speakers on course-related topics

Approximate length: one page.

How to Write a Business Letter

As a business letter writer, you face two general concerns:

- ☐ How to organize the letter into its parts according to a particular format
- ☐ How to tailor your message to your reader for the best results

Let's look at the letter's form first because this will be the framework for your message. As you follow the guidelines shown in this section, check the example letters at the end of the unit for illustration.

The Parts of a Business Letter. In order, from the top of the page to the bottom, there are about eight parts to a business letter (some are optional).

1. The heading
   - **a.** Your address
   - **b.** The date

Note: When you write on company letterhead paper, the address is printed on the sheet; enter the date only.

2. Inside address
   - **a.** Receiver's name and title or department
   - **b.** Receiver's/company's address

   Note: When you don't know the receiver's name, use titles and departments that suggest your purpose, such as "Personnel Manager" and "Accounting Department."
3. Subject line
   - **a.** States topic of letter in a brief phrase
   - **b.** An *optional* part
4. Salutation, or greeting
   - **a.** "Dear Mr./Ms.______:" (last name only)
   - **b.** "Dear Field Service Manager:"
   - **c.** "Dear Customer Accounts Department:"

   Note: If you don't know the person, don't assume a male reader; that is, don't use "Dear Sir:"; use options like *b* and *c* instead.
5. Body of letter
   - **a.** First paragraph: state your purpose
   - **b.** Middle paragraphs: explain or support your purpose
   - **c.** Last paragraph: promote action or good will
6. The close: "Sincerely," "Yours truly," or the like
7. Your name
   - **a.** Signature first
   - **b.** Your typed name below the signature
   - **c.** Your title if you are writing on letterhead paper
8. Enclosures and copies section
   - **a.** Call attention to attached documents such as checks, copies of bills, a resume, and similar items
   - **b.** Write "Enclosure(s)," name the attachment(s), or give the number of attachments
   - **c.** After the enclosures notation, if any, list people who are getting copies of the letter, if any.

Formats. The sequence of letter parts is given in the preceding list, but you still have to decide how to *space* the parts on the page. A schematic layout showing the spacing of letter parts is presented in Figure 3–7 for the block letter format. The best way to learn formats is to look at the example letters at the end of this unit and the next. These illustrate the three major letters formats in detail.

1. *Block format* (see Figure 3–8)
   - **a.** All letter parts are at the left margin
   - **b.** Double space between paragraphs

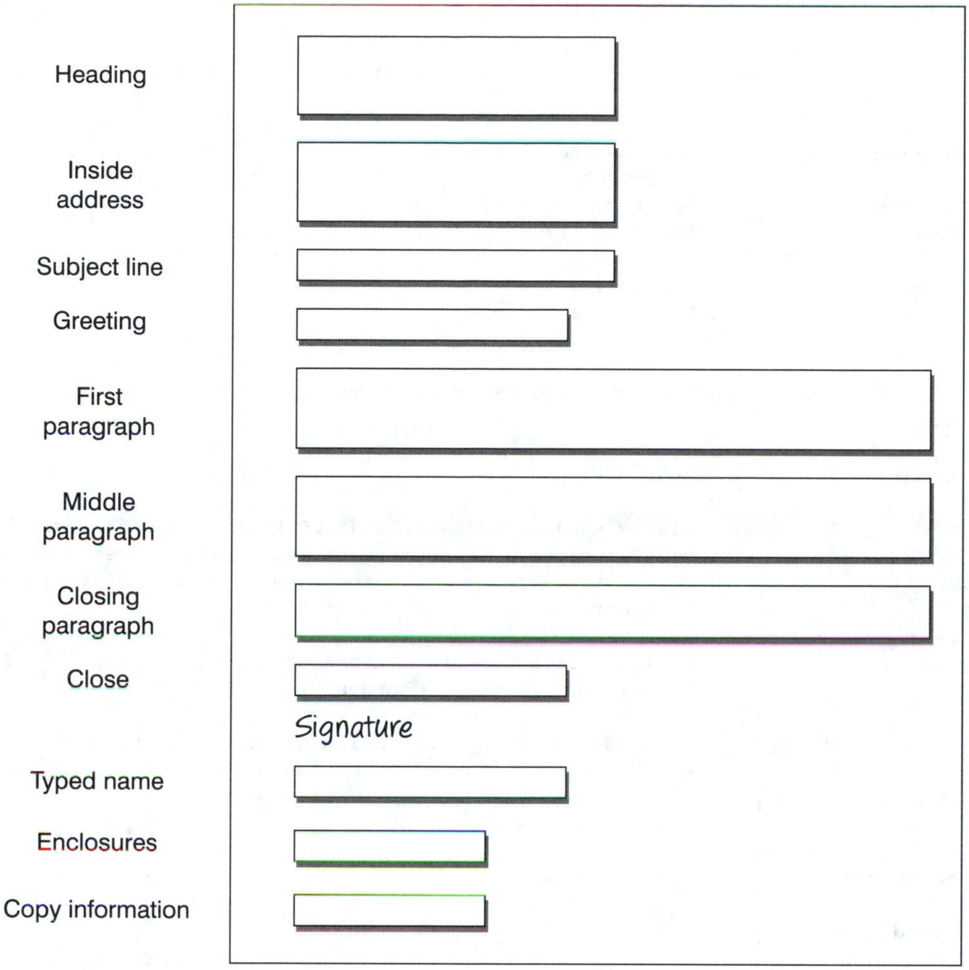

Heading

Inside address

Subject line

Greeting

First paragraph

Middle paragraph

Closing paragraph

Close

Signature

Typed name

Enclosures

Copy information

FIGURE 3–7
The spacing of letter parts in schematic outline; the block format is illustrated.

   - **c.** Double space between most letter parts
   - **d.** Extra space between the two addresses and below the close (for your signature)
2. *Semiblock format* (see Figure 3–9)
   - **a.** Most letter parts are at the left margin
   - **b.** Heading and signature sections at the right margin
   - **c.** Extra space for your signature
3. *Indented format* (see Figure 3–10)
   - **a.** Like the semiblock style, except . . .
   - **b.** Paragraphs are indented

Heading	453 East Main Street Apartment 1-E
Address **Date**	Decatur, IL 93102 April 14, 1996
Inside **Address**	Ace Audio, Inc. 859 Fordham Avenue Westlake, NY 20940
Subject **Line**	Subject: Ace DX899 CD Player
Greeting	Dear Ace:
Body	(1) Please send me one Ace DX899 CD Player unit for use with my JVC L-A31 amplifier. According to your customer representative Mr. Walsh, these units are compatible. Please confirm this with your shipment. Enclosed is a personal check for $253.75, (2) including $8.75 for (3) parcel post shipping charges. (3) I would appreciate your mailing the CD Player by the end of this month. Thanks (4) for your attention.
Close	Yours truly,
Signature	Sandra Mason
Typed Name	Sandra Mason
Enclosures	Enclosure: Check

1. Name of product
   a. exact name
   b. quantity
   c. model/catalog number
   d. technical specifications
2. Price and method of payment
3. Shipping method and date (if important)
4. Thanks

FIGURE 3–8
An order letter in block format.

1235 Adams St.
Porter, WI 54321
October 11, 1996

Customer Service Department
Zenax Corporation
1010 Fair Drive
Omni, MN 43210

Dear Service Manager:

(1) I'm interested in buying a hard disk drive unit. Based on the catalog description, your 8 megabyte Z199-5 sounds like a good buy. However, I need to know the following:

(2)
1. Can it interface directly with my Ajax 386 personal computer, Model 2000?
2. What is the length of the warranty of this unit for 386 machines?

(3) I'd appreciate a reply by November 7. I need the unit for a school lab project that is due November 21. (4) Thank you for your help.

Sincerely,

Alexander Scott

Alexander Scott

(5)
c: Bill Farlane
Audrey Spellman
Terry Wright

1. Reason for inquiry
2. What you want to know
3. Desired notification time
4. Thanks
5. Copies to lab partners

[If request is to a person, not a company, include a stamped, self-addressed envelope].

FIGURE 3–9
An inquiry letter in semiblock format.

17 Terminal Street
Apartment 4-J
Farnsworth, VA 42391
March 28, 1997

Discovery Credit Card, Inc.
187 Greene Street
Washington, MO 53879

Subject: Unauthorized Charges, Account 25-498-1094

Dear Customer Service Manager:

(1) Enclosed is a copy of my bill dated 25 March for Discovery Card Account No. 25-498-1094. In addition to seven charges that I made, the bill also includes the following charges:

3/10	Hair Stylistics	Farnsworth, VA	43.50
3/11	Groovy Disco	Allenbee, VA	127.35
3/11	Holiday Inn	Allenbee, VA	58.65

I did not make these three charges. (2) As you can see from the 3/15 transaction of $250 to the Hyatt Regency in Chicago—I was in Chicago from 3/09 to 3/15. I have included (3) a copy of that hotel receipt as verification.

(4) I would appreciate your deducting these amounts from my bill. A check for $465.15 is included to cover the seven charges I did make this month, but excluding the $229.50 in unauthorized charges.

(4) I do not know how these erroneous charges occurred. I have kept in my possession the only card I was issued, and have not allowed anyone else to use it. Please let me know what should be done to prevent further unauthorized charges.

Yours truly,

Ellwood P. David

Ellwood P. David

(5) Enclosures

1. Identify transaction specifically: names, dates, places
2. Explain problem
3. Enclose proofs: bills, receipts, canceled checks
4. Request action to fix problem
5. Enclosures of Discovery bill, hotel receipt, check

FIGURE 3–10
A complaint letter in indented format.

Tailoring the Message to Your Reader. Picture your reader, and shape your message to her needs. Whether you know the person or not, you're writing in a business situation, and that means the reader is busy and subject to many demands. You'll get a better response if you help the reader help you.

1. Get right to the point—in the first sentence if possible.
2. Keep your message short and simple.
3. Make it easy for the reader to understand you.
   a. Give complete information.
   b. Be precise: Refer to model numbers, dates, prices, and other particulars.

These suggestions for tailoring your message to the reader are practical applications of the "you attitude" in business communication. In this approach, the reader's needs are placed before the writer's, and the message is shaped to express the reader's point of view. In some cases, a "you" attitude is expressed by revising I-we-us statements into you-yours statements:

We: We stock the following models . . .
You: You may order the following models . . .
We: We received your order . . .
You: Thanks for your order . . .
We: I don't recall your order . . .
You: Please help me trace your order by . . .

Your style and tone in a business letter should emulate face-to-face communication. That is, write naturally and avoid excessive formality. If you wouldn't use stiff language and pompous words in conversation, why put them into your letter? Instead, be courteous but friendly.

If your letter is a complaint and you don't feel very friendly, remember that a positive tone is more likely to gain solutions for your problem than a negative, accusing one. Compare:

Your clerk didn't complete the order form properly, so we can't send you the equipment you wanted.

versus

Please fill in the power requirements section on the order form so we can ship the proper unit to you as soon as possible.

Be polite at least. The person who reads your letter is probably not the cause of your problem. In any case, your purpose will best be served by a natural and clear statement in a courteous way (e.g., see letter in Figure 3–10).

Good-News vs. Bad-News Letters. It is worth learning the most effective approaches for at least two basic kinds of business letters—those that convey a positive message or response and those that bring disappointment or refusal. While all letters should employ the strategies discussed in the previous section, the good-news and bad-news varieties can be further shaped to respond to the reader.

A *good-news* letter presents the main point first, and follows it with explanation or clarifying detail (see Figure 3–11). A *bad-news* letter, on the other hand, begins with a *buffer* that cushions the negative news, then presents reasons for the refusal (while emphasizing the "you attitude"), and ends with a counter offer (see Figure 3–12).

As you look at Figures 3–11 and 3–12 consider how the good-news/bad-news strategies are being applied in letters written by a company to a professional society student chapter requesting a speaker for its technology forum.

Examples of Letters

Use the letters in Figures 3–8 to 3–12 as models for your own business communications.

UNIT 3.4 RESUMES AND APPLICATION LETTERS

One of the key opportunities to use your writing skills will come in the job-search process at the end of your academic career. You will need to develop two documents for this purpose: (1) a resume and (2) a cover letter requesting an interview. These documents should be a matched pair working smoothly together, each in its own way reinforcing your goal of arousing the interest of employers and encouraging them to call you for an interview.

This unit will show you how to develop each of these documents for maximum effectiveness. The example resumes and letters are also matched in content so you can see how they reinforce each other.

The Resume

A resume is a one-page advertisement designed to interest employers in giving you a job interview. It presents your qualifications in a concise, user-friendly way, stressing the skills, education, and experience that you believe will appeal to a limited group of employers. You might make 50 to 100 copies of your resume, for example, and mail it to companies in certain product or service categories, to those in an area where you would like to live, to those with good growth records, or to those with a combination of such factors.

You might also consider creating a "home page" for yourself on the World Wide Web, and including your resume within this set of materials. See the section on "Putting Your Resume on the Web," on page 164.

Meanwhile, the more you know about prospective employers, the better you can tailor your resume to their needs. A resume is a general document, but it is not aimed at all employers. In other words, you will need to do some company research before you plan your job campaign.

Resume Formats. Three types of resumes are common:

1. *Chronological* (see Figure 3–13)
   - **a.** Lists education and employment history

ABCo 2835 W. Melbourne Way Longhorne, TX 75675 214-928-6700

March 14, 1996

Eugene A. Stanton
Events Coordinator
Instrument Society of America
Dallas Area Student Chapter
4800 Carpenter Freeway
Irving, TX 75065

Dear Mr. Stanton:

We have asked Dr. Elvina T. Whatley, Vice President for Research and Development, to join your Technology Forum April 23, and she has accepted with pleasure. Both Dr. Whatley and I were extremely impressed by the thoughtful planning you have put into this event, and we are confident of its success.

Since Dr. Whatley lives in the area, she will not be needing hotel accommodations. However, she has asked that you call her during the following week at her office (X 2807) to discuss arrangements for projection equipment and other matters related to the presentation.

With respect to the honorarium cited in your letter, Dr. Whatley has asked that it be donated in her name to the Irving Symphony Orchestra during its annual fund drive in May.

We wish you success in your program and other chapter endeavors.

Sincerely,

Richard Goldsmith
Human Resources Manager

c: Dr. Elvina Whatley

FIGURE 3–11.
A good-news response letter.

 ABCo 2835 W. Melbourne Way Longhorne, TX 75675 214-928-6700

March 14, 1996

Eugene A. Stanton
Events Coordinator
Instrument Society of America
Dallas Area Student Chapter
4800 Carpenter Freeway
Irving, TX 75065

Dear Mr. Stanton:

Your Technology Forum is one of the most impressive programs I've seen in my 15 years in this position. Dr. Elvina Whatley, our Vice President of Research and Development, asked me to convey her congratulations as well. She feels your program should attract a large audience and bring you favorable notice in the local media.

Dr. Whatley would have been delighted to accept your invitation to join the forum. Unfortunately, she and two of her colleagues, both of whom would also have been suitable participants, will be in Riga, Latvia, the week of your forum and so will be unable to attend. At Dr. Whatley's suggestion, I would ask that you contact Dr. Burundi A. Rohani of Camden Systems Inc. per the enclosed business card. Dr. Rohani is a noted authority in the field of digital signal processing and should make an excellent alternative speaker.

If your chapter members are interested, we would be pleased to invite you to tour our facilities this summer, preferably during June or July, and to talk to our research staff. Give me a call and let's explore the idea further.

Best wishes for the success of the forum.

Sincerely,

Richard Goldsmith
Human Resources Manager

c: Dr. Elvina Whatley

FIGURE 3–12.
A bad-news response letter.

   - **b.** Reverses the order of occurrence
2. *Functional* (see Figure 3–14)
   - **a.** Lists major qualifications and skills
   - **b.** Uses education and employment data to illustrate skills
3. *Combined* (see Figure 3–15)
   - **a.** Uses a chronological format
   - **b.** Includes some functional elements

What to Include. Some information is essential, some is optional, and some should be omitted. Omit personal information that could be the basis of possible discrimination.

Essential Information.

1. Name, address, and phone (with area code)
2. Education (in reverse order)
   - **a.** All schooling/training *after* high school
   - **b.** List degree/certificate received; if none, give the years you attended
   - **c.** School name, city, state
   - **d.** Relevant courses and activities
3. Work experience (in reverse order)
   - **a.** Job title; duties and accomplishments (use action verbs: "built," "repaired," "supervised," "organized," etc.)
   - **b.** Dates worked, company name, city, and state
4. Qualifications
   - **a.** Not jobs you've had, but what you *know* and *can do*
   - **b.** Skills, knowledge, licenses, ability to use equipment, and so on
5. Accomplishments
   - **a.** Not what you can do, but what you've *done* that reflects hard work, creativity, leadership, reliability, and similar qualities
   - **b.** Proofs: awards, citations, high grades, publications, language certificates
6. Military service (if any)
   - **a.** Branch, highest rank, honorable discharge
   - **b.** Job duties, security clearances
   - **c.** Schools attended, academic achievements and recognitions
7. Citizenship, if other than this country

Optional Information.

1. Employment or career objective: tells what kind of position you want now or in the long run
2. High school: include only if courses or activities relate to your career field
3. Clubs and organizations, including offices held
4. Job-related interests and hobbies
5. Statement that references are "available on request"

Omitted Information.

1. Personal data: age, sex, height, weight, health, handicaps, your picture
2. Street addresses and phone numbers of schools and jobs
3. Names of supervisors
4. Salaries
5. Reasons for leaving prior jobs
6. Names and addresses of references (to protect their privacy)

Writing and Layout.

1. List only things that sell you as a good employee. Don't list everything.
2. Make the resume easy to read. Employers skim many resumes rapidly.
   a. Items are briefly stated
   b. Sections are labeled and separated
   c. Most important selling points stand out
   d. Page is not cluttered or dense with print
   e. Blank spaces guide reader and ease the eye
3. Professional appearance
   a. Neat
   b. Well written and error free
   c. Typed, typeset, or printed by a high-resolution printer
4. Distinctive and individualized: Make yours stand out from the hundreds employers read.

Examples of Resumes. Chronological, functional, and combined resumes are shown in Figures 3–13, 3–14, and 3–15, respectively.

Putting Your Resume on the Web. The World Wide Web is a network of Internet computers with special hypermedia programs that permit use of text, graphics, sound, real-time video, and multimedia combinations. The hypermedia capability of these computers allows creation of documents that contain embedded references to other pages of information as well as images and sounds. These other resources may reside in the same or different computers, all of which are linked within the Web. When a user clicks on a highlighted word or phrase, for example, she is able to access information or illustrations from the same computer or from elsewhere (see Figure 3–16).

A primary feature of the Web is the *home page*—a compilation of information formatted as a hypermedia document. The information may be about an organization or an individual and is commonly much more extensive than the contents of a traditional printed page. The "page" concept is expressed on the Web in a covering list of topics similar to a table of contents or outline. These topics may then be further explored on subsequent "pages". A home page is identified by a Uniform Resource Locator (URL) code such as the following sequence: http://www.arizstate.edu/ftp/progs/.

For individuals, a home page may serve as a self-presentation or introduction to people with similar interests; for many professional people, it is also an opportu-

FRANCIS R. LENNON

102 Newhaven Drive
Selma, Alabama 50030
(515) 297-9022
flennon@sand.el.edu

Or contact through
Sandhurst Placement Office
(515) 343-9765

EDUCATION

1993–1995 *Associate of Applied Science in Electronics*
Sandhurst Technical Institute, Selma, Alabama
- financed 80% of education through part-time employment

1991–1992 Priory Community College, Selma, Alabama
- 18 credit hours in mathematics and liberal arts

October 1991 *Blueprint Reading Certificate*
Horry-Georgetown Technical Institute, Conway, South Carolina

June 1987 *Color Television Service Certificate*
Radio Electronics and Television Service, Conway, South Carolina

MILITARY

1987–1991 *Avionics Specialist* (F-16), United States Army
- Honorable discharge
- Trained air crew in communication/navigation devices
- Maintained, repaired aircraft communication/navigation systems
- Secret security clearance

EXPERIENCE

1992–Present *Electronics Technician*, Masterwork Systems, Selma, Alabama
- Repair Hitachi and Conrac Video Monitors
- Maintain and repair laser printers
- Repair computer motherboards

1991–1992 *Crew Leader*, Wendy's, Conway, South Carolina
- Promoted to Crew Leader after 1 month
- Responsible for 6 employees
- Trained new employees

1980–1987 *Farmhand*, family-owned farm, Conway, South Carolina
- Repaired and operated heavy machinery
- Veterinary care of animals

QUALIFICATIONS

Electronics
- Troubleshooting circuits
- Reading schematics
- Soldering
- Function generators
- Power supplies
- Oscilloscopes

Carpentry — All power tools
Painting — 10 years' experience
Auto Work — Motor maintenance, repair, body work

ACHIEVEMENTS

- Designed and built remote-controlled robot in high school
- Learned welding and did spot welding in automotive shops
- Designed and constructed voice-band PCM system

FIGURE 3–13
Chronological resume.

ROBIN BURROWS

2494 Packwood Circle #5
Los Angeles, CA 90015

Office: (714) 787-4756
Home: (714) 787-6549
E-mail: rburrows@ucla.bus.edu

OBJECTIVE

A position with opportunity for advancement using skills in PUBLIC RELATIONS, MARKETING, REAL ESTATE, and GRAPHICS DESIGN

QUALIFICATIONS

Management
- office manager
- make all bank transactions
- procure office equipment
- authored office policy manual
- property manager for 3 duplexes
- manage office procedures for sales agents

Bookkeeping
- full-charge bookkeeper
- manual and computer accounting
- converted books to computer processing
- payroll/quarterly tax reporting
- workers' compensation accounting
- record keeping for condominium projects

Organization
- organize new condominium projects
- liaison between client and contractor
- arrange FNMA approval

Advertising/ Public Relations
- layout and design of:
  —ad copy
  —brochures
  —new client welcome package
- wrote PR proposal for state tourism division

License
- California Real Estate License

EXPERIENCE

1992–present *Office Manager*
Bascom Real Estate Co., Los Angeles, CA

EDUCATION

May 1992 *Bachelor of Arts in Advertising and Public Relations*
University of California at Los Angeles

FIGURE 3–14
Functional resume.

MARTIN C. MORAN

Current:
2770 Northeast Expressway
St. Louis, MO 63125
(314) 987-6511

Permanent:
102 Palms Drive
Fort Walton Beach, FL 33942
(815) 376-1624

GOAL

To secure an entry-level programmer/analyst position

EDUCATION

BACHELOR OF SCIENCE IN COMPUTER INFORMATION SYSTEMS, December 1995
Washington University, St. Louis, Missouri

ACADEMIC DIPLOMA, June 1989
Foster Williams High School, Fort Walton Beach, Florida
- Secretary, Future Business Leaders of America

ACHIEVEMENTS
Hard-working
Determined
Creative

3.5/4.0 GPA
Dean's List
Worked 34-40 hours while attending school
Developed own graphics software

SKILLS

Languages	*Applications*	*Business*
• DB2	• General Ledger	• Cost Accounting
• COBOL	• Report Generation	• Managerial Accounting
• C++	• Accounts Receivable	• Economics
• FOCUS	• Accounts Payable	• Marketing
• BASIC	• Inventory Control	• Business Law
• OS/JCL	• Master File Update	• Small Business Systems

EXPERIENCE

Supervisor

ASSISTANT MANAGER, Night Lites, St. Louis, 1993–present
—deposit $3000–$5000 nightly
—hire and schedule staff
—supervise evening business

Problem Solver

BARTENDER, Elbow Benders, St. Louis, 1992–1993
—increased sales volume 10%, lowered liquor costs 3%

Responsible

MANAGER, Mama Rosa's Pizza, Fort Walton Beach, Florida, 1991–1992
—ran entire operation of successful food establishment
—supervised 10 employees
—missed only three days in two years

Team Worker

HEAD WAITER, Steak and Ale, Fort Walton Beach, Florida, 1989–1991
—trained and coordinated wait staff

INTERESTS

Repair and tune automobiles, design computer circuitry

REFERENCES

Available upon request

FIGURE 3–15
Combined resume.

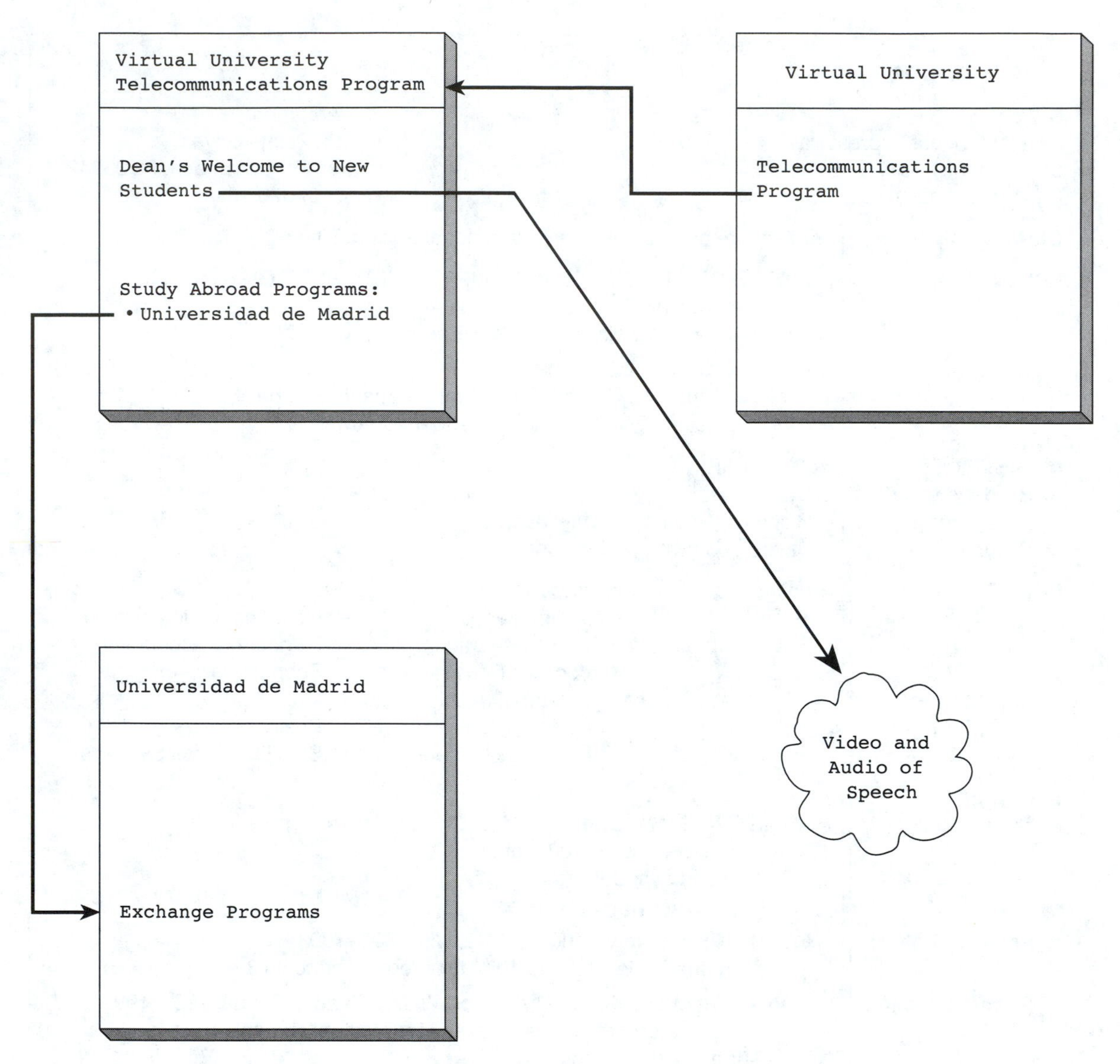

FIGURE 3–16
A hypermedia document on the World Wide Web allows users to access stored audio, video, and text, either in the same computer or elsewhere on the Web.

nity to present their capabilities and credentials to potential employers. Such people may wish to include a resume within their home page.

Authoring a Web document such as a home page requires use of *hypertext markup language* (HTML), a computer language that provides specifications for formatting information. For example, if you wanted to begin your document with the title "Resume," you would give the following instructions: <TITLE>Resume</TITLE>. Fortunately, HTML is a relatively easily learned method that allows beginners to quickly author documents with the aid of one of the many available guides.[1] Furthermore, software is becoming available that automatically converts documents into HTML code.

In designing your resume for presentation on the Web, use the same principles you would apply to a paper resume. Keep in mind, however, that a Web page resume is a presentation to an extremely broad and varied audience. While you might focus a resume in certain ways when you send it to a selected group of employers, a Web resume casts a much wider net and should not foreclose opportunities through excessive focusing. Differences in focus between a Web resume and a targeted resume may show up in the statement of career objectives, for example, and in the selection of details from prior employment experiences.

Once you have created a Web resume, you should register it with the appropriate research services (e.g., Netscape) so it is available to Web searchers (see page 210). College placement or departmental offices may also sponsor lists of Web resumes for their students.

The Application Letter

When you mail a resume to an employer, always introduce yourself and your qualifications with an application letter. A good application letter is essential to gaining more attention for your resume than for the many others the employer receives.

The resume is a general description of your qualifications, but the letter that covers it (or *cover letter*) is a personal, customized message to a specific person in a specific organization. The letter will be effective if it shows how you can benefit the company. By relating your skills to a company's needs, you show that you really want to work for this organization, not just find any job. To convey this interest, you must first research the company to learn its needs.

If you are replying to a published ad, you have some of the necessary information. For more information on companies, including those that haven't advertised openings, check the business indexes in your library or school placement office. The Moody's manuals, Peterson's guides, Thomas Registers, and other reports offer company and product descriptions that you can use for this purpose. (See Unit 5.1, Research, for a list of indexes.)

Types of Application Letters. In planning a job-search campaign, you are most likely to consider writing the following types of letters:

1. *Letter of Invite:* written to a company that has advertised an opening. (See Figure 3–17.)
2. *Letter of Prospect:* written to a company not advertising an opening; the wise candidate doesn't wait for want ads but writes prospect letters. (See Figure 3–18.)

---

1. See, for example, Laura Lemay's, *Teach Yourself Web Publishing With HTML in a Week* (Indianapolis: Sams Publishing, 1995) or Mary E. S. Morris's, *HTML for Fun and Profit* (Mountain View, CA: SunSoft Press, 1995).

3. *Broadcast Letter:* a "form letter" with no reference to the company it is sent to; you can cover more companies this way but not as effectively as with prospect and invite letters. (See Figure 3–19.)

Essential Content

1. State immediately why you're writing, that is, "to apply for the [state specific] position."
2. Explain why you want to work for this particular company (except in a broadcast letter).
   a. Good reasons: type of work, growth opportunities, challenge
   b. Bad reasons: lots of time off, strong retirement plan, "easy work"
   c. Inappropriate reason: salary—because this is a negotiable figure
3. Show why they should hire you. Don't just repeat your resume but summarize and highlight key strengths in three main areas:
   a. Education
   b. Work experience
   c. Personal attributes such as leadership, ability to get along with people, and your strong motivation to succeed

   Note: Many letters claim desirable personal qualities. The effective ones prove their point by citing offices held, promotions achieved, supervisory responsibilities received, and so on.
4. State when you can begin work.
5. Request an interview.
6. List phone numbers at which you can be reached during work hours.
7. Thank the reader for considering your application.

Writing and Appearance

1. Keep the letter direct and brief. If you can't do it in a page, chances are you've lost the reader anyway.
2. Write in natural English that avoids excessive formality and stiffness.
3. Use a word processor to customize the letter for a specific person or department at a specific company. Call the company to get the name, title, and department if you have to.
4. Use a word processor to customize the letter to reflect different strategies and emphasize different benefits as appropriate.
5. Make sure the letter is error free: *no* spelling errors, *no* usage errors, *no* grammar mistakes, or the reader may think, "If you're such a hot prospect, how is it you didn't take time to proofread something as important as an application letter?" And remember, the surest way to get your application rejected is to misspell the reader's name or the name of the company.

Examples of Letters. Figure 3–17 shows a letter of invite; Figure 3–18, a letter of prospect; and Figure 3–19 a broadcast letter.

102 Newhaven Drive
Selma, Alabama 50030
January 16, 1996

Mr. Alexander Simpson
Personnel Department
Marvell Communications, Inc.
140 Petulia Street
Selma, Alabama 50032

Dear Mr. Simpson:

I am applying for the entry-level electronics technician position with Marvell Communications, as advertised in the *Selma Times* on January 15. I am very interested in your plans to develop low-cost home satellite dishes, and I feel I can contribute to your goals in this area. (1) (2)

In February I will receive an Associate of Applied Science in Electronics degree from Sandhurst Technical Institute. In this program I have studied communications extensively, as well as circuit analysis/troubleshooting and digital devices. I also have 450 hours of hands-on lab experience in both system and component level electronics. In my part-time position at Masterwork Systems, I repair video monitors, laser printers, and computer motherboards. My merit reviews state that I am an industrious, responsible worker; I would like to put my skills to work for your company. (3)

I would be able to start work on February 17, and would welcome a chance to talk with one of your representatives about the opening. You can reach me at (515) 297-9022 or through the Sandhurst Placement Office at (515) 343-9765. Thank you for your consideration. (4) (5) (6) (7)

Yours truly,

Francis R. Lennon

Francis R. Lennon

1. Apply for the position
2. Your interest in the company
3. Your qualifications
4. Availability for work
5. Request for interview
6. Phone numbers
7. Thanks

FIGURE 3–17
Letter of invite.

2494 Packwood Circle #5
Los Angeles, CA 90015
September 8, 1996

Ms. Joan Seabury
President
Seabury Realty Company
250 Tanglewood Avenue
Los Angeles, CA 90017

Dear Ms. Seabury:

As office manager of Bascom Real Estate Company for the past four years, I have noted the success and widespread respect that Seabury Realty has earned in the Los Angeles area. Bascom agents who deal with Seabury agents in closing home sales often comment on their professionalism and dedication. I would like to be part of such a team, and am therefore submitting my credentials for your consideration. (1) (2)

In my role as office manager, I have developed skills in management, bookkeeping, organization, and advertising/public relations. I am particularly proud of the office policy manual I wrote and the books that I converted to computer processing. I also created a well-received public relations campaign for the California Division of Tourism. I think I can contribute significantly to your organization. (3)

If a position requiring my talents becomes available, please consider the enclosed resume. I would be available on two weeks notice. At this time, I would welcome the opportunity to tell you my ideas on improving office efficiency. If you would like to get in touch, I can be reached at work at (714) 787-4756 or at home at (714) 787-6549. Thanks for your attention to my credentials. (4) (5) (6) (7)

Sincerely yours,

Robin Burrows

Robin Burrows

1. Your interest in the company
2. Apply for the position
3. Your qualifications
4. Availability for work
5. Request for interview
6. Phone numbers
7. Thanks

FIGURE 3–18
Letter of prospect.

2770 Northeast Expressway
St. Louis, MO 63125
May 15, 1996

Mr. Samuel Baker
Personnel Director
Topspeed Information Systems
633 Laclede Station Road
St. Louis, MO 62967

Dear Mr. Baker:

As a recent college graduate majoring in computer information systems, I am eager to begin my career as a programmer or systems analyst trainee for a respected company that needs an energetic, team-oriented worker. I would like to present my credentials for your consideration. 1

I am prepared to become a valuable employee not merely because of my schooling, but also by six years of work experience. Last December, I received a Bachelor of Science degree in Computer Information Systems from Washington University. My studies focused on the business uses of computer programming and systems analysis. I learned six different languages, including the 4th generation FOCUS; with such wide exposure, I believe I can pick up new languages rapidly. I created programs for ten standard business applications, including Accounts Receivable and Payable, Inventory Control, and Master File Update. I also took many business courses, including managerial accounting and marketing. Moreover, my work record for several employers indicates that I have learned to 2

- work well with others
- handle responsibility
- hire, train, and supervise employees

I hope these qualifications interest you, should you need a programmer/analyst with good potential for professional growth. I am available to begin work on two weeks' notice, and would be pleased to come in for an interview at your convenience. You can reach me at (314) 987-6511 weekday afternoons. Thank you for your consideration; I hope to hear from you. 3 4 5 6

Sincerely yours,

Martin C. Moran

Martin C. Moran

1. Apply for position
2. Qualifications
3. Availability for work
4. Request for interview
5. Phone number
6. Thanks

FIGURE 3–19
Broadcast letter.

Unit 3.4 Exercise

Professional Development Project: A Portfolio. The preceding discussion of resumes and job application letters looks at strategies for presenting yourself effectively to potential employers. A key recommendation is that you think of ways to *prove* that you are an outstanding candidate, and not just claim it. To help you do this, consider developing a professional portfolio.

Portfolios have been adopted by many colleges as part of their educational process to promote self-exploration, learning, assessment, and other purposes. A professional portfolio is a more focused device related to personal development and achievement, particularly in the context of professional and career goals. Such a portfolio is a collection of work that you select and arrange during your academic career for presentation to potential employers at the time of graduation. Among the kinds of materials that may be included in the portfolio are the following:

- ☐ An updated set of resumes arranged in different ways to highlight education, experience, and attributes
- ☐ Memberships in professional societies
- ☐ Honors, awards, prizes
- ☐ Recommendation letters
- ☐ Reports
- ☐ Lab reports
- ☐ Research papers
- ☐ Computer projects
- ☐ Technical design projects
- ☐ Creative works
- ☐ Documented problem-solving experiences
- ☐ Photographs and diagrams of work done
- ☐ Videos

To an employer, an effective portfolio shows not only your academic and job-related achievements but your ability to plan and organize your self-presentation. For best results, start early in your academic career and be deliberate about saving work you do and feedback you receive.

Among the ways you might arrange your materials are (a) chronologically, (b) by types of materials, (c) by employment objectives, or (d) in "developmental order," showing growth or progress. You can also get advice about how to organize the portfolio from instructors and from staff in the Career Services or Dean of Students office.

SECTION 4

Oral Communication

This section draws together a series of units focused on the practical needs of spoken communication. The speaking situations addressed here include group presentations and meetings as well as one-on-one conversation. Media, including the telephone and various audio-visuals, are brought in because they introduce special advantages and requirements. Emphasis is also given to the other end of the communication chain—listening.

The subjects discussed here are all improvable skills. They require an understanding of the speaking situations to which they may be applied. They require conscious effort and preparation. Then they must be practiced.

They are also "portable" skills. If you train yourself to listen effectively, you will hear and understand more in the classroom and in work or social situations, whether you are listening as an audience of one or many.

These units also point to the similarities and differences between speaking and writing. Many of the principles of good communication are the same for the two forms. There are similarities in the organization and development of reports, for example, whether these are spoken or written. On the other hand, you must also understand the strengths and limitations of each form of communication and give these due attention.

Instructors sometimes ask students to write reports and then to prepare brief oral summaries of their papers. This dual-track approach is also common in industry. To the extent that you practice delivering the same message in writing and speech, you can improve *both* skills.

UNIT 4.1 ACTIVE LISTENING

Although hearing is a natural process, *listening* is not. Listening, however, is a skill that can be learned and improved with practice. Active listening requires concentration and effort on your part to make you a full participant in the communication process. Though you are not speaking, your mind and observation should be fully engaged, and your critical sense fully deployed.

At work, listening is an important part of the job requirements. You must not only grasp what supervisors, colleagues, and customers are telling you, but you must understand the emotions, priorities, and pressures that may lie behind the words. Listen to the body language and tone of voice as well as the words, and respond to what a person feels, along with what he says.

Responding is a key part of actively listening on the job. By sending signals back about how you are receiving the message, you can help the speaker make a point and assure her that you understand and emphasize. You may not always agree, but you can let the speaker know you are listening by interjecting, "I see . . . " or an equivalent comment, and by looking directly at him.

Also-learn how to formulate questions

In the classroom, the lecture method puts its own burden on you to remain an active participant. The lecture is an efficient way to cover material, but it also tends to encourage passivity. Even if you are taking notes, it is far too easy to become an inactive, uncritical recorder of what the instructor is saying. You may write everything down, only to discover later that your scribbles make little sense.

To become a good listener in the classroom, you have to accept the fact that active listening is hard work. It demands as much of you as lecturing does of the instructor. Being a student is truly an occupation (see Figure 4–1).

FIGURE 4–1
The key to active listening is to work at processing the incoming message: translate, select, relate, evaluate.

Active Listening Strategies

The following suggestions can help you get the most out of the time you spend in class.

1. Don't allow distractions to interfere with your listening. Focus on the lecture: Tune out squeaking chairs, coughing and whispering, paper shuffling.
2. Listen to what the instructor says, and don't be distracted by personal traits such as an accent or a physical mannerism.
3. Do listen to tone and inflection, however. Most instructors reflect a natural enthusiasm for what they consider especially important.
4. Listen also for indicators like "the main points are" or "the most important difference is," and to things the instructor repeats for emphasis. Mark these points in your notes for review.
5. Summarize and think about what is being said while you listen. You can think as much as three times faster than anyone can speak, so there's plenty of time to organize your thoughts for note taking, formulate questions, or even decide how you feel about what's being said.
6. Try to relate what you hear to what you already know. You will understand and remember a point better if you can relate it to previous lectures, to lab and homework, and to the material in your text. In fact, the instructor will assume that you have been keeping up in all these areas.
7. Approach lectures with an open mind. Be ready to take in new ideas and information. Remember, the instructor has a message that is worthwhile for you.

Unit 4.1 Exercise

1. Write a one-paragraph summary of part of a lecture. Check your version against a summary prepared by the instructor.
2. Draw a picture or diagram of something described by the instructor (e.g., a mechanism, a graph, a network, a flowchart). Compare your version to a copy provided by the instructor.
3. Practice active listening with a friend, roommate, or classmate. Take notes on an explanation or description. Then restate your understanding of the message in your own words.
4. Apply a *triad* approach to the situation in Question 3. Bring in a third person to listen to the statement and the restatement. Let the third person comment on the accuracy of the restatement. Then discuss the differences among the three points of view.
5. Get two classmates to watch a presidential address or other major television presentation with you. Afterward, give each person three minutes to summarize the main points of the address. Discuss differences among the summaries.
6. Use a VCR to tape the TV presentation in Question 5. Refer to the tape in comparing and contrasting summaries.

7. Try the parlor game called "Telephone." Whisper a message to one person, who relays it to the next, and so on. The last person restates the message aloud and compares it to the original. In this case, don't worry about word-for-word restatement, but check whether the main ideas were communicated (see Figure 4–2).

UNIT 4.2 NOTE TAKING

Note taking is a skill with many applications beyond the classroom. You're likely to need it to organize your work, to keep track of specific information, and to remind yourself of due dates and commitments. Still, the classroom is the place where good note-taking skills are most needed at this stage of your career. This unit is a guide to developing these skills so you can get the most out of the challenging task of being a student.

Note-Taking Strategies

Probably the most important aspect of note taking is *effective listening*. Don't let the words of the instructor go directly to your notebook without logging them into your brain on the way. Instead, let your notes do their real job: remind you of what you already know.

Notes should support the work you do in class—the work of listening actively, incorporating information into your mind, and relating what you hear to what you have learned before. Your notes should be in the form of outlines, phrases, or just key words. They should represent important ideas or major points that will trigger your memory of the full subject.

FIGURE 4–2
"Telephone" shows the results of poor transmission and poor listening.

Class Notes. When you take notes, put them into your own words. The mental process of rewording what your instructor says helps you to learn the material as you encounter it. Also, don't try to record every word. Use your judgment about what's important to remember. Listen for cues like "The key points here are . . . " and "Be sure to distinguish. . . . " In these cases, the instructor is alerting you to important material.

Ask questions. If you don't understand something or it doesn't jibe with what you read in the text, ask your instructor about it, either right then or after class. Your main purpose in class is to learn *at the moment*—not to take notes from which you plan to learn later.

If your mind wanders or you miss a point for any reason, leave some room in your notes and ask another student or the instructor to help you fill the gap after class. If the point is essential to your understanding the rest of the lecture, don't hesitate to stop the instructor and ask right away.

After class, review your notes. Make sure you've identified the subject the notes cover, the day they were taken, and the part of the textbook they relate to. Add a summary at the end of your notes if you like. It could clarify a lecture that got off the topic or lacked organization.

Reading Notes. When taking notes on reading assignments, many of these same points apply. Your main goal is to *read effectively* and learn the material as you read it. Taking notes as you read will make reviewing for tests easier. It can also provide many of the notes you would take in class later. In class, you can follow your reading notes and underline points your instructor emphasizes, and you can fill in extra notes next to the originals.

Notes on your reading assignments can best be taken in outline form using the textbook's own organization. You can use titles and subtitles to create the skeleton of the outline before you begin reading. Then fill the important points into the outline as you read.

Depending on how closely your instructor's lectures follow the text, you might develop a system that blends both sources into one evolving set of notes, as shown in Figure 4–3.

UNIT 4.3 TELEPHONE CALLS

Much important business is conducted over the telephone. But telephone communication doesn't always go smoothly, for both human and technical reasons. You may have trouble getting through to the right person, hearing and being heard, understanding or being understood. You can overcome many of these problems if you understand the common pitfalls of business calling and practice making calls in ways that avoid potential difficulties. Preparation and practice will also help you gain confidence in a skill that is basic to business and professional life.

READING NOTES	CLASS NOTES
Microcomputer impact (pp. 177–89)	Wed. 12/8
A. Encourages entrepreneurial spirit	A2. Also for better customer service
1. Distributes processing capabilities	
2. Processing at point of transactions	
B. Changes responsibilities	B. Decline of centralized systems
C. Risks—loss of control over data/info	department
	Executive/secretary role reversal
	Exec keyboards, sec'y edits
D. Advantage: personal productivity	D. Balance against (C) – overlap
E. Planning and forecasting	and duplication re information
1. Spreadsheets	
2. Data base software	
F. Others	F. Also, local area networks for
1. Electronic mail, lists	info sharing
2. Graphics	* More on E next class
	* Stress here on changing roles
	and responsibilities

FIGURE 4–3
Two columns of notes on the same subject.

SUGGESTED CALLS

Use the telephone report form in Figure 4–4 to make calls such as the following:

- ❑ Request information on products and services from a company.
- ❑ Request information on specific equipment, parts, or software.
- ❑ Request catalogs, brochures, data sheets, and similar items.
- ❑ Request general information on career opportunities (types of positions, entry requirements, company locations, etc.). *Note:* You may need to clarify that you are *not* job prospecting at the moment.

Telephone Report Form

(Use for each separate call, except to directory assistance numbers.)

Purpose of call: ____

____

Who was called: ____

Telephone number (and extension): ______________________________

How number was obtained: ______________________________________

____

Results of call: ____

____

____

____

____

Notes on problems encountered: _________________________________

____

____

____

____

____

____

____

____

____

____

FIGURE 4–4
Telephone report form.

- ❑ Leave a voice-mail message stating your purpose and giving a name and a number where you can be reached.

How to Make Telephone Calls

We all know how to use the telephone . . . or do we? Have you ever had trouble explaining your purpose over the phone or understanding someone else's response? Have you ever called a large organization and been transferred from person to person, having to explain your purpose over and over?

Using some of the basic telephone techniques will help make your calls more pleasant and productive.

1. *Speak into the receiver.* As simple as this sounds, it's commonly forgotten. And don't be afraid to suggest this to another person you are having trouble hearing. ("Excuse me, but I can't hear you very well. Are you speaking into the receiver?")

2. *Speak clearly and slowly.* Without the benefit of facial expressions and gestures, it takes a little longer to get ideas across. Clear speech and expression become that much more important.

3. *Before you dial, plan what you're going to say.* This technique gives callers more confidence and saves time by helping to avoid false starts. ("Trans-Global. Can I help you?" "Oh. Umm. I'd like to know if you . . . well, I need some information about . . . about a . . . ")

4. *Before you dial, decide on a likely person or department to ask for.* Many organizations have switchboard operators who answer incoming calls and transfer them according to the caller's request. These operators work with lists of names, departments, and functions, but their information is not extensive, and they don't know everything about an organization. It's up to you to place your request in a general category when you don't know a specific name. ("Hello. I'd like to speak to someone in your parts department.")

5. *Hold the specifics of your call until you've been transferred to the right person,* or you may face the frustrations shown in Figure 4–5.

6. *Ask for names.* Whenever possible, ask operators for the name of the person who would handle a request in your general category. Then ask to speak to that person. ("Hello, operator. Could you tell me who can give me some information on your tutorial software?" "That would be Alice Andrews." "Thank you. Could you connect me please?") Also, ask for the name of everyone you are transferred to. This information will help you avoid getting the same person again and save time if you have to call back.

7. *When you get detailed information, write it down, and ask the other person to verify it.* ("So that's Model X4T for $14.95 per 100 in lots over 500, right?")

FIGURE 4–5
When you call a business organization, make sure you are transferred to the person who can help you before you launch into a full explanation of the problem.

These tips, of course, apply only after you have a number to call. To get names and numbers of organizations when you only know what you want, try these two resources:

1. *The yellow pages.* This telephone book lists organizations under categories of service or product they supply. There are two separate books—the B book, for retail organizations that deal with the general public, and the C book, for industrial and wholesale suppliers and services. Both books are available free from the telephone company.

2. *Magazines and newspapers.* Many ads list telephone numbers and addresses for organizations. Large companies frequently provide toll-free numbers. All numbers with an 800 area code are toll-free.

If you have a person's or organization's name but no number, try these guides:

1. *The white pages.* Residences and organizations are listed in alphabetical order. (Some organizations or individuals don't list themselves here, however.)
2. *Directory assistance.* Dial 411 to get numbers and area codes.
3. *Long-distance directory assistance.* Dial 1-area code-555-1212.

After you find the number and decide what you're going to say, you may still encounter frustration. Besides switchboard operators—who are very busy, sometimes not well informed, and who *can* make mistakes—you may get receptionists, secretaries, and co-workers who answer at the extensions you're connected to.

Not everyone takes the time to be helpful. Persistence and patience will be your most valuable assets in this case. If you know what you want and state it clearly each time, you should eventually get the information you're after.

Dealing With Voice Mail

Today many organizations, as well as individuals, use voice mail to handle incoming calls. These computerized telephone answering systems can serve as automated switchboards, directing calls to various destinations and allowing callers to record detailed messages. Messages are stored until they are accessed and may then be saved, deleted, or sent to another extension.

These systems have certainly reduced the extent of "telephone tag," but they are not without pitfalls. Callers to organizations may find themselves frustrated by the endless options they must respond to as the program sorts them through predesigned categories at its own deliberate pace. Messages may not be returned as expected by those who are away for extended periods but have not alerted callers. By the same token, receivers of voice mail may find messages incomprehensible or lacking vital information.

As a caller, you can improve your chances of navigating voice mail if you (a) obtain and use extension numbers to get through the external "switchboards" of organizations, (b) speak slowly and clearly in recording your message, (c) give the date and time that you are calling, (d) state your purpose and give your name at the beginning of the message, and (e) repeat your name and give your telephone number—slowly—at the end of the message.

Example of a Successful Call

Read the following example and then plan some business calls that will not only give you practice but also deliver some useful information as well. Use the form in Figure 4–4 if appropriate.

MIKE CALLS TRANSGLOBAL

Switchboard operator: Good morning! TransGlobal!

Mike: Good morning. I'd like to speak to someone who could give me information on your new wireless laptop computers.

Switchboard operator: That would be our Research and Development department. Shall I connect you?

Mike: Yes, please. Thank you. [CLICK!]

R&D: Research and Development!

Mike: Yes, hello. This is Mike Khan. Who am I speaking to please?

R&D: I'm Donna Myers. Can I help you?

Mike: Yes, Ms. Myers. I would like some information on your new wireless laptops. I'm working on a school project at Madison Technical College.

Donna Myers: Well, let's see. I think Pat Levine could help you. Hang on. [CLICK!]

Voice: "Hello. . . . This is Pat Levine. . . . "

Mike: Hello, Ms. Levine. I'd . . .

Voice: " . . . I'm not at my desk right now, but leave a message at the tone and I'll get back to you." [Beep!]

Mike: Oh, hello. This is Mike Khan, Ms. Levine. It's now 8:30 on Monday morning. I need to get some information on a school project. I'll call back in a half-hour.

[30 minutes later]

Switchboard operator: Good morning! TransGlobal!

Mike: Good morning. May I speak to Pat Levine, please?

Switchboard operator: Thank you. [CLICK!]

Unknown: Chris Marlowe!

Mike: Oh, I'm sorry. I asked to speak to Pat Levine. Do I have the right extension?

Chris Marlowe: Nope. Hang on . . . umm . . . okay, Pat's on 415. Hold on, I'll transfer you. [CLICK!]

Pat Levine: R&D Department. Pat Levine speaking.

Mike: Hi, Mr. Levine. My name is Mike Khan, and Donna Myers suggested you might be able to give me some information on your new wireless laptop computers. I'm a student at Madison Technical College, and I need to evaluate several models for a school project.

Pat Levine: What is it you wanted to know?

Mike: Specifically, I'm trying to make an estimate of their potential for field service applications. Does yours . . .

UNIT 4.4 MEETINGS

Collaborative and group approaches have become more common in the workplace in recent years, largely for competitive reasons. Technical professionals may spend as much as 25 to 50% of their time in department meetings, task forces, or committees. Much of this activity places a premium on being a good participant or contributor, rather than on leading the meeting.

And while there are definite expectations of a meeting leader, the role of an effective participant is less clear. Yet, when meetings are held to solve problems or generate ideas, members are expected to contribute significantly. An understanding of how such task groups operate can help you formulate strategies for participation that respond to the challenge.

Meeting Dynamics

Most task-oriented meetings are informal gatherings of small groups. A limited size—perhaps three to eight persons—helps the group manage time, hear from the participants, and reach closure on issues. Use of smaller groups also places more responsibility on each member to carry the load during meetings. That's the tricky part.

How do you do your part without stepping on other people's toes, or, conversely, without letting them down when they expect ideas or solutions from you? Does your role call for pressing your own ideas or supporting the ideas of others? And where should you draw the line between supporting others' ideas and giving them honest feedback?

It may help to think first about the "rules" for conducting a meeting of this kind. Despite the informal tone, there is usually an underlying structure defined by an agenda or stated purpose. You can ordinarily rely on the meeting leader to introduce and pursue this agenda: to lead the discussion and manage time. Let the meeting leader also set the rules of the discussion: the order of speakers and whether speakers will be called on or will volunteer their comments at appropriate moments.

Expect the meeting leader also to summarize the proceedings so that participants go away agreeing on what happened. And expect the leader to conclude the meeting by telling you what the next steps, if any, will be. Such conclusions may inform group members that their suggestions will be adopted, or passed on to others, or that they should follow up on unresolved issues for a future meeting.

Effective Participation

One of the best ways you can ensure a maximum contribution at the meeting is to prepare. Review the agenda, study materials passed out beforehand, gather up materials of your own. Because you are preparing to participate, formulate questions you wish to ask or points you want to make when the time comes.

Come to the meeting with a positive, expectant attitude. Expect to contribute your ideas, but expect to learn from others as well.

Your own ideas will gain a better reception if they are well thought out, perhaps pretested, during the preparation before the meeting. Of particular value will be data or other evidence supporting your proposals. (For an extreme example, see Figure 4–6.)

Present your ideas in general terms but back them up with specific instances and examples to help others see their usefulness. If possible, choose applications that are meaningful to several different colleagues to earn their support.

Be prepared also for a lukewarm or indefinite reception. Don't be discouraged. There will be opportunities to revise your ideas and perhaps introduce them in other situations. Take a long view of the process.

When others speak, give them the benefit of your active listening response. Let them formulate their points thoroughly and help them clarify points when necessary. Give them feedback on their ideas.

If your reaction is mixed, comment on the positive aspects first before you raise a question about other parts. Invite the group to comment on the questionable parts as well, to focus their collective wisdom on the problem.

FIGURE 4–6
You can make a valuable contribution to any meeting if you prepare thoroughly and back up your ideas with solid evidence.

Remember that in a meeting you must balance the duties of generating ideas and of responding to others' ideas. Both actions are expected of you, and both make a contribution.

Teleconferences

Advances in communications technology have made it easier for people from different locations to hold meetings. Long-distance telephone conference calls have become an increasingly used alternative to having participants travel to central sites. In the more advanced form of *video*conferencing, participants are not only able to hear but also see one another and view documents and models. Video, however, is less common because of the added expense, greater complexity, and continuing glitches, such as the lags between voice and picture.

You can maximize your effectiveness in a teleconference by observing a few principles of good participation. Let the convener of the conference set the procedural ground rules; for example, will participants be called on in turn, or may comments be offered at any time? If the ground rules are not given at the start of the call, ask for clarification.

Remember also that people from dispersed sites may not know each other by voice or name, so precede each comment with identifiers: "Don, this is Sharon Jones from Atlanta Two. . . . " Check that others can hear you (by asking if they can), speak into the audio device, speak clearly, and be explicit enough to make up for the lack of visual cues such as gestures and facial expressions.

Unit 4.4 Exercise

1. Set up an informal meeting of a study group to prepare for a major exam. Limit the group to about five, and assign each member a specific subject area to work up before the session. As each member presents the assigned topic at the meeting, all group members join in a constructive but critically engaged discussion. The goal of preparatory work and discussions is to maximize each member's score on the exam.
2. Your instructor may wish to set up a series of project groups to work on a major assignment due at the end of the term. The assignment might be one that includes research, writing, and oral presentation. Possible approaches include the following:
   - **a.** Groups are self-selected but may be filled out by the instructor.
   - **b.** Groups choose a leader and divide the parts of the project among themselves.
   - **c.** Groups set up a project schedule with milestones for completion of parts.
   - **d.** Groups meet periodically to review progress and help solve problems.
   - **e.** Groups earn a "base" grade for their work, but each member's grade is adjusted according to a weighting system that includes the instructor's observations, class evaluations, and ratings of one another's contributions to the project by team members.

UNIT 4.5 ORAL REPORTS

Project teams, customer contacts, and presentations to management all require good oral communication skills. The practical need is for brief, focused presentations, usually to small audiences. Such presentations should be well prepared but not given as memorized "speeches."

The oral reports you give in class will increase your confidence and ability to think on your feet. In the longer term, your personal presentation skills will help you not only in launching a career, but also in career advancement.

POSSIBLE ASSIGNMENTS

- ❑ Summarize a class lecture in a 2-minute presentation.
- ❑ Explain the objective of a lab assignment—before the rest of the class attempts the lab.
- ❑ Compare and contrast two or more personal computers on word processing, spreadsheet, or data base capabilities.
- ❑ Report progress on a particular project.
- ❑ Present a summary of a formal lab report.
- ❑ Give a detailed explanation of a particular process, device, or program.
- ❑ Present information on a company in a selected career field.
- ❑ Give a persuasive presentation showing that a particular system, type of equipment, or procedure is the best for a certain application.
- ❑ Give an oral presentation of your part of a written formal report on a team project.
- ❑ Evaluate the presentations of classmates. Use a prepared evaluation form such as the one shown in Figure 4–7.

Approximate length: 1 to 15 minutes.

How to Prepare and Deliver Oral Reports

In content and structure, a prepared oral report is quite similar to a written report. Both presentations typically need an *introduction,* a *body,* and an *ending.* But additional factors must be considered in each of these sections when a report is presented orally.

Introduction. The introduction not only should announce your subject, purpose, and scope but also *gain the audience's attention.* Attention-getting devices include

Name of Speaker ______ Class/Section ______ Date ______

In each of the following areas, assign a number indicating how you would rate the speaker's performance:

4 = VERY GOOD   2= FAIR
3 = GOOD   1 = NEEDS MUCH IMPROVEMENT

Note: If an item does not apply, write in NA.

- ❑ Did the speaker capture your interest and attention right from the start? ______
- ❑ Did the speaker effectively introduce the report by explaining the purpose of the presentation? ______
- ❑ Did the speaker use a logical plan of organization that led naturally to the conclusion? ______
- ❑ Were charts, transparencies, models, or handouts well prepared and smoothly worked into the presentation? ______
- ❑ Did the speaker end the report effectively by summarizing the main points, presenting conclusions, making recommendations, or otherwise finding an appropriate ending? ______
- ❑ Did the speaker use a natural delivery, with voice and gestures suited to the subject? ______
- ❑ Did the speaker use good grammar and word choice? ______

TOTAL SCORE ______

COMMENTS: ______

Evaluated by: ______

FIGURE 4–7
Evaluation form for the oral report.

raising a thought-provoking question, making a dramatic statement, or presenting an interesting fact related to the central point of the presentation: "According to a recent *Wall Street Journal* article, seven out of ten college graduates will experience the obsolescence of the job in which they began their careers. . . . " You can also refer to the audience directly or to some common experience of theirs to catch their interest: "I see that at least some of you survived the trip downtown this morning. As survivors, you will appreciate . . . "

On the other hand, introductions shouldn't be overly clever, startling, or gimmicky. Such tactics may gain attention, but not necessarily in a favorable way. They also make transition into the presentation itself more difficult because of the gap between your actual purpose and the introductory shenanigans.

Body. In preparing this section, you can use many of the strategies applicable to written reports. (Review the organization and development methods suggested in the Introduction and in Unit 2.4, Formal Reports.)

Oral presentation imposes its own special requirements as well. One of these is the need to limit the number of points you cover. In a written report, the reader can review earlier material as necessary, but in the oral presentation, all points must be held in the audience's mind during the delivery. So keep the number of main points to about five or less. And because of the normal limitations of the attention span, you should explain each point in more detail, give more examples, and orient the audience more (e.g., "A *third* way of dealing with this problem is to . . . ").

A good rule of thumb for structuring the body of an oral presentation is to build in as many signposts as you can. Don't worry about orienting your audience too much. Instead:

1. Tell them what you're going to tell them.
2. Tell them.
3. Tell them you've told them.

Ending. Audience interest is highest at the beginning and end of a presentation. You can therefore use your ending to summarize or highlight the main points with the assurance that they will get a better-than-average hearing. As in the written report, use a method of ending that is most suitable to your subject and purpose: a summary, a conclusion, a recommendation, an appeal to action, or an appropriate generalization.

A *summary* may be best for an informative speech making a number of points. You might recap the major functions of a new device, for example, so that others could decide whether it compared favorably with existing technology. If your purpose was to evaluate the feasibility of the device, on the other hand, you might emphasize your *conclusions:* "This equipment can do more things for us faster and at lower cost than the present system. . . . " In this case, the appropriate final word would be a *recommendation:* "For these reasons, I believe we should begin to phase out the present system and replace it with digital signal processing technology at the earliest opportunity."

An *appeal to action* would be an appropriate ending for a persuasive presentation. A heightened appeal might be used when the need is great, as in a famine relief drive: “Please make a pledge today. Tomorrow may be too late, literally.” A *generalization* or perhaps an appropriate *quote* might be an effective way to restate a central point of the presentation in these or other cases.

Many variants of these endings can be used to fit the circumstances, as long as they leave the audience thinking about your main points and purposes. There are bad endings as well:

- ☐ Don’t end with an apology (e.g., “I’m sorry there wasn’t time to cover all the topics? . . . ”).
- ☐ Don’t ramble on beyond a logical stopping point.
- ☐ Don’t introduce new material (e.g., “And by the way, it occurs to me that . . . ”)
- ☐ Don’t change the mood you have established (e.g., by telling a joke at the end of your famine relief speech).

Presentation

At formal occasions such as professional association meetings, it is customary to simply read a written report aloud. In most other cases, a reading would clash with audience expectations of a more personal approach. A more active presentation method is to use note cards that contain only key words or phrases instead of complete sentences. The words or phrases are designed to trigger the speaker’s memory of a specific portion of the report, but the exact words are chosen in the delivery. Cards with specific quotes, facts, or figures can be included and read from directly as they come up in the presentation (see Figure 4–8).

For many people, giving a speech is a scary prospect. You may also feel nervous as you plan your presentation. In such cases, there are some practical ways to reduce your anxiety and get through the initial stages of your talk.

- ☐ Preparation is the best antidote. By learning your subject thoroughly and preparing a general plan of presentation along with specific details, you can assure yourself that you know more than the audience about this topic and have something to offer them. Your confidence should increase.
- ☐ Practice is the natural partner of solid preparation. Practice your delivery to gauge timing, and try your material on a willing friend for feedback.
- ☐ Visit the site of your presentation, if possible, to check the layout of the room for audiovisual needs, and for the psychological assurance created by a familiar environment.
- ☐ Take a few deep breaths (on your way to the front of the room perhaps), and pause before you begin. Your first words will have more “voice” behind them and you’ll have time to gather your thoughts or arrange materials.
- ☐ Concentrate on friendly faces in the audience at first to build up confidence.
- ☐ Move your arms or feet, point to something, or otherwise gesture to release nervous energy.

FIGURE 4–8
Speaking from notes can help you achieve a more natural delivery that better meets audience expectations.

For the presentation as a whole, use the following strategies to get the most out of your planning and preparation:

1. Look at your audience and talk directly to them. (If faces make you nervous, look at *foreheads* instead. It will appear to the audience that you are looking right into their eyes.)
2. Stand on both feet. (This will stop nervous shifting and make it easier to move in any direction and to use gestures.)
3. Let your arms move to accompany your words (as they would normally in conversation). Don't tie your arms down by putting your hands in your pockets.

4. Speak loudly enough so everyone can hear you. (Aim your voice at the people in the last row.)
5. Speak clearly.
6. Speak slowly enough to give your audience time to absorb what you say and to let you breathe naturally. (This will also help you deal with stage fright.)
7. Talk naturally, but use your preparation to select *precise* terms.
8. Let your voice show that you're interested in your subject. (Let it rise and fall naturally as you emphasize certain points.)

Unit 4.5 Exercise

1. Rachel is planning her presentation on the use of multimedia technology in the classroom to a group of faculty and deans at Madison Technical College. She sketches out several possible *introductions.* Which of the following have the most potential in your opinion?
   - **a.** It's safe to say that the days of "chalk and talk" are behind us—or are they?
   - **b.** Tenure for everyone! Now that I have your undivided attention, let me tell you about . . .
   - **c.** Today, you can bring "Star Wars" into the classroom to help students learn.
   - **d.** How many of you are from this area? (Show of hands.) Good, good. This is my birthplace, too, so I guess we have something in common.
   - **e.** How many here can say that your instructional methods include strategies to promote active learning on the part of your students?
2. The *body* of Rachel's presentation on multimedia technology has two main parts: an outline of primary capabilities and identification of equipment needs. In the section on necessary equipment, Rachel plans to talk about the following items:
   - ☐ A microcomputer with an EGA/VGA graphics card
   - ☐ A standard overhead projector
   - ☐ A computer-screen reader (that mounts on the overhead projector)
   - ☐ A remote control for the computer
   - ☐ Wireless response pads (that allow students to key in numerical responses to instructor questions)

   Sketch out the transitions for Rachel's presentation of this section: (a) what should she say at the beginning and end of it, (b) what parts should she establish, and (c) how should she start and end these parts? Compare your scheme with others in your class.
3. Rachel's purpose in making the presentation is to persuade Madison Technical College to invest in the multimedia software produced by her company, HyperExtend Inc. Which of the following *endings* would you suggest she use, and why?
   - **a.** In conclusion, our multimedia services can deliver 12 main kinds of enhancements to ordinary classroom presentations. These are . . .
   - **b.** I hope I've shown you that HyperExtend offers greater potential for interactivity than any other multimedia service on the market today.

c. I'm sorry I forgot to bring the laser pointer and the hand-held remote to show you, but I hope you got something out of the rest of the presentation.
d. Well, as Franklin Delano Roosevelt once said, "We have nothing to fear but fear itself."
e. If you think this technology might have some potential for your college, I'd like to set up some meetings between your instructors and our software designers to develop a few prototype packages.

UNIT 4.6 VISUALS FOR ORAL PRESENTATIONS

Almost any oral presentation can be enhanced by visual materials. If the audience can see what is being presented, as well as hear it, their understanding and recall of the material will be greater.

From the speaker's standpoint, visuals help present difficult or complicated material clearly, make routine material livelier, and ease the pressure to "perform." To gain these advantages, *plan* your use of visuals and then *coordinate* them with what you say.

Planning for Visuals

1. Analyze your speech and its audience. Review the purpose of your speech, what you intend to say, and the background of your audience.
   a. What needs to be clarified? Look for complex, technical, intricate material; relationships between variables; patterns of change; actions hard to describe in words; ideas new to the audience.
   b. Where does the speech need to be broken up or livened up? Look for sections dense with information, full of abstractions and generalities, or drawn out with procedural detail.
   c. What needs to be emphasized? Look for key information and evidence. Consider your purpose again, and highlight elements central to it. Give your conclusions or recommendations special emphasis.
2. Determine where to add visuals.
   a. Identify the specific places for visuals. In most cases, show the visual after presenting the idea in words first. This keeps attention on your words as the primary agent of presentation.
   b. Anticipate the impact of visuals on the flow of your speech. If you plan to write or draw on the board, for example, it will be more difficult to keep talking meanwhile. The flow will also be interrupted if you need to switch off the lights, draw shades, walk over to a display table. Look for natural breaks in your speech.
3. Select the most effective combination of visuals. Aim for at least two types of visuals in a short speech and at least three types in a longer one. In your selection, consider the combinations of factors presented in matrix form in Table 4–1.

TABLE 4–1
Options in Selecting Visual Aids

	Tables	Charts/ Graphs	Pictures	Words/ Phrases	Physical Appearance	Functioning
Flipchart/stand	•	•		•		
Posterboard/easel	•	•		•		
Chalkboard	•	•		•		
Handouts • Multiple copies • Single copies	•	•	•	•		
Displays • Real objects • Scale models					•	•
Demonstrations • Equipment/materials • People					•	•
Projections • Slides/films/videos • Transparencies • Of computer screens	•	•		•	•	•

Designing Visuals

Many benefits of visual aids can be lost through poorly designed or prepared materials. Here are five characteristics you should work for in your visual aids:

- ☐ Large scale
- ☐ One main idea
- ☐ Immediate impact
- ☐ Drafted appearance
- ☐ Highlight the idea

Make sure the visual is *large enough* to be seen from the back of the room. Check this yourself if possible. If you're not sure how large the room will be, err on the side of making your item larger than necessary. For visuals of a limited size, such as transparencies, use all the available space, handletter labels rather than typing them, or just switch to another format (posterboard, chalkboard, etc.) that allows larger size. Remember that everything else you do will be defeated if the audience can't see your work.

Restrict each visual to one *main* idea. If you want to illustrate a series of related things or a process, break it up for separate presentation. One of your visuals may summarize the entire series or process, but this should be followed by separate views of the parts (see Figure 4–9). Remember that your visuals are meant to clarify ideas and to focus attention. Crowded visuals are little better than the dense verbiage they are supposed to illuminate.

Make visuals clear and easy to *read at a glance,* so that the message almost leaps out at the viewer. Simplify and focus the idea. A company's recent financial performance, for example, should be reduced to simple terms that reflect a trend. Notice also how the effectiveness of the chart in Figure 4–10 is enhanced by clear (and visible) labeling of the axes.

To simplify ideas, you might produce a list of key words or phrases to serve as cues for the explanation. Thus, if you are discussing the duties of business managers,

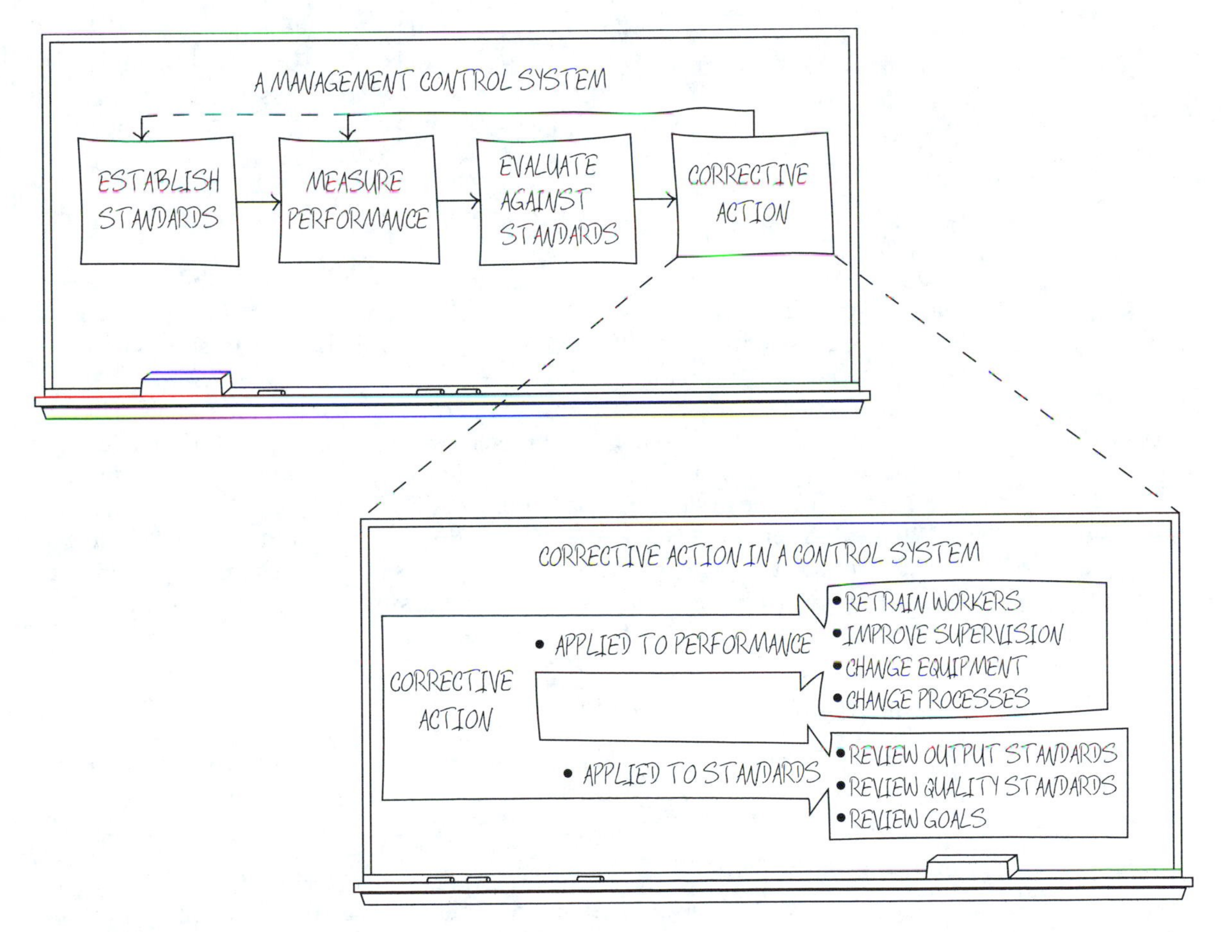

FIGURE 4–9
A visual showing one main idea broken down into two main parts and clearly defined subparts.

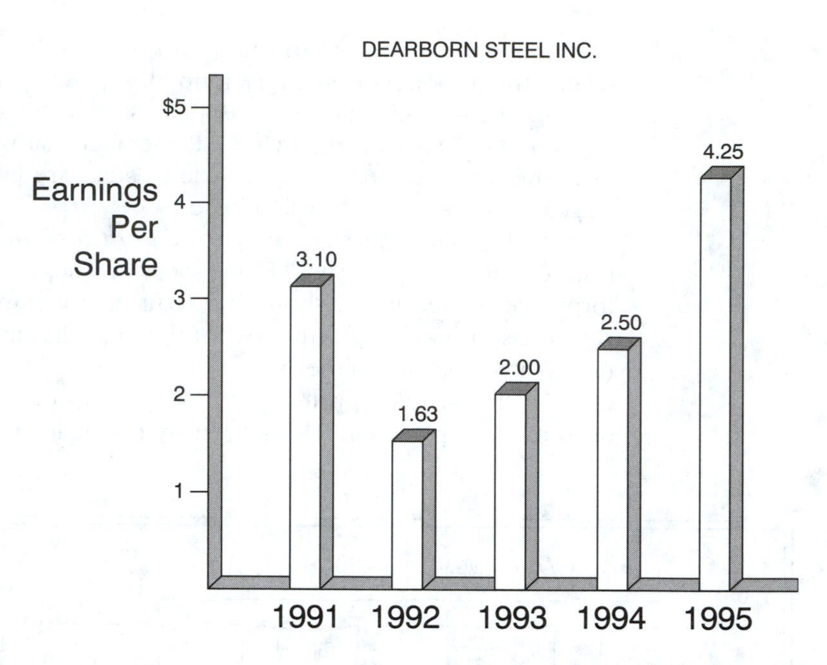

FIGURE 4–10
Financial performance at a glance. Note the simplified design, absence of detail, and clearly labeled axes. Letters and numbers are oversized for visibility.

you may introduce a visual that boils these complex responsibilities down to their essentials (see Figure 4–11).

Draft your visuals so they are neat, correctly proportioned, and professional looking. Sketch a preliminary version and then use drafting aids such as straight edges, curves, and templates. Where possible, use several colors to strengthen contrast and emphasize key points.

Look for ways to *highlight* or *dramatize* the material in your visuals. This will help you gain attention and interest, and earn a better reception for your ideas. If you have some artistic ability, you might be able to put your ideas into visually striking terms much as the political cartoonist or the graphic artist does.

Whatever your artistic talents, simpler options are usually available. Simple geometric figures such as pyramids, triangles, rectangles, and circles can highlight ideas. A classic example is the pie chart, which dramatizes the relationships between parts and the whole. See Figure 4–12 for an example.

Showing Visuals

Once you have planned and designed effective visuals, you need to make sure you give them a good showing and work them smoothly into your speech.

Preparation Before the Speech

1. Set up the visuals and equipment you will use (e.g., flip chart and stand, slide projector, VCR, model on display table). Cover the visuals so they don't steal attention from your words.

2. Sort out the accessories you will need (e.g., chalk, eraser, marking pens, pointer).

3. Arrange for necessary assistance, perhaps from members of the audience (e.g., to turn off the lights, draw the shades, pass around handouts, or assist with a demonstration).

Handling Visuals During the Speech

1. Position visuals so they are visible from all parts of the room. If possible, a central and *high* location is best. The screen for an overhead projector, on the other hand, should be to one side so the machine does not get in the audience's way. In some cases, you may need to pan visuals slowly across the audience's field of vision.

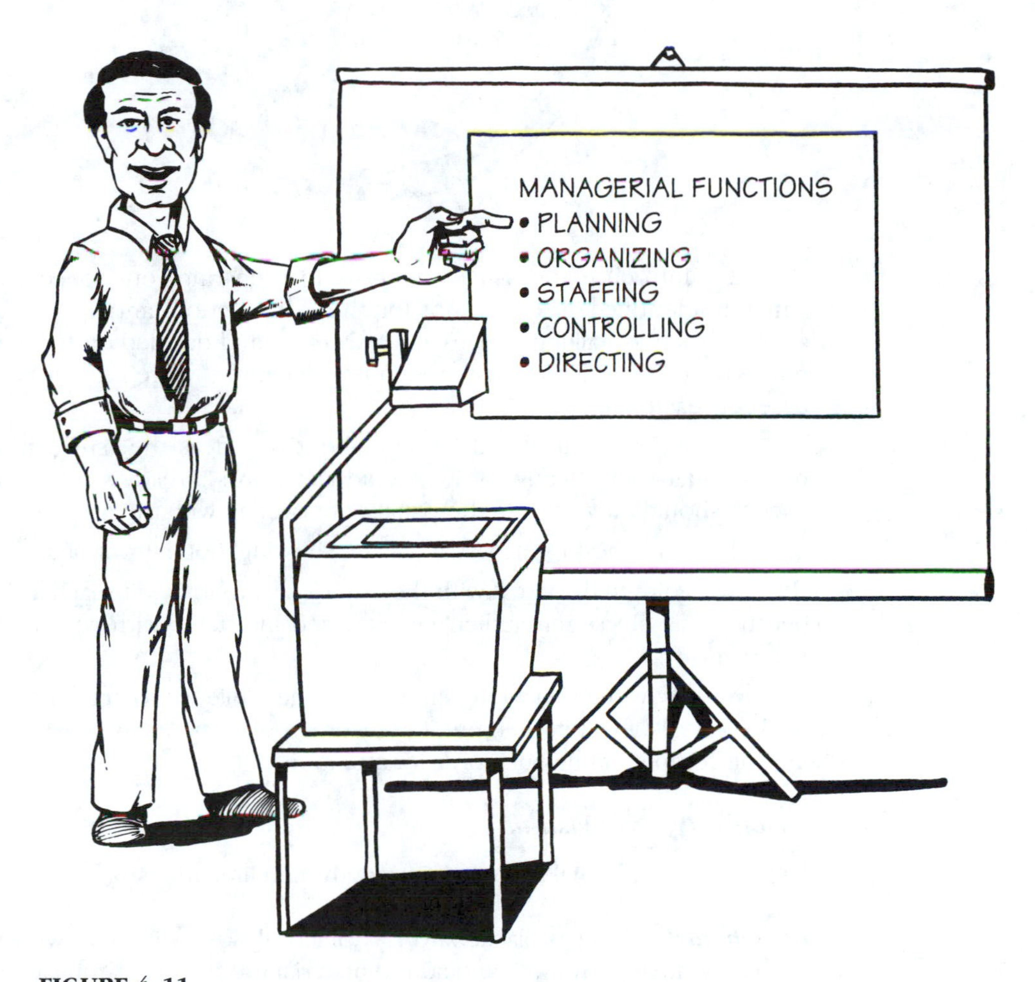

FIGURE 4–11
An overhead transparency listing key words that summarize a more detailed discussion by the speaker.

FIGURE 4–12
A pie chart that highlights a financial issue. TravelCanada's controller might ask why Toronto accounted for nearly half the expenses, given its central location, or why Calgary's bill was so low. Were bookings proportionate?

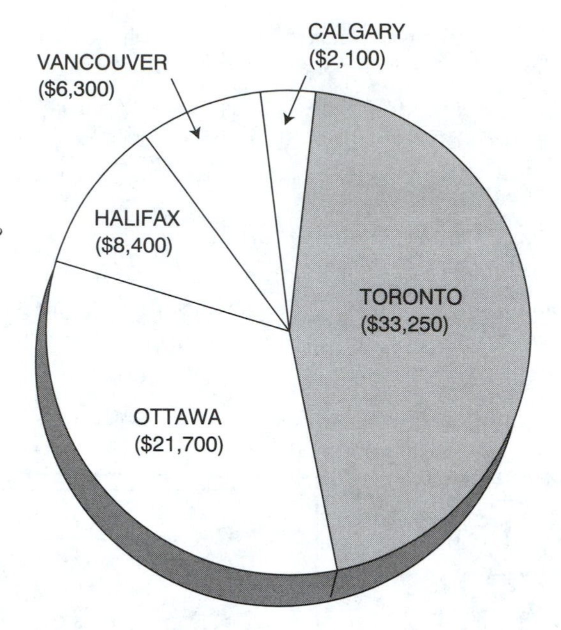

2. Introduce any visual you show. In preparing your speech, you have presumably identified an exact spot for the introduction, anticipated any necessary actions (such as walking over to an easel, etc.), and decided on the actual phrase of reference (" . . . which is illustrated in this diagram"; "Here is a cut-away view of the breeder reactor.").

3. Explain your visuals or talk about the ideas they suggest. Don't merely repeat what they already show or read the words already written on them. Your speech should *interact* with the visuals, not overlap with them.

4. Stand next to the visual you are showing, not in front of it.

5. Point to the visual with the arm closest to it. Don't reach across your body, because this blocks the audience's view and interferes with your eye contact (see Figure 4–13).

6. Show the visual long enough for the audience to grasp its details. Show it while you talk about it, then put it away or cover it up. An attractive visual left standing will compete with your words for attention.

Specific Types of Visuals

Each category of visual aids has its strengths and limitations.

Blackboard. Using a blackboard is a practical way to interact with the audience. You can write or draw on it and add to material progressively. You can erase once the point has been made. It's probably best for words and simple diagrams; more complex graphics are difficult and time-consuming.

Flip Chart. The flip chart is a "portable blackboard," equally good for interaction and progressive development. It may be brought close to the audience and allows you to uncover prepared work and instantly remove a used visual. Several colors can be added. Compared to the blackboard, the flip chart offers less space to work with.

Posterboard. A posterboard is ideal for prepared material, such as dramatic visuals in multiple colors. Posters can be mounted on an easel, a flip chart stand, the blackboard's chalk tray, or supported by the speaker.

Photographs. Photographs may be used as an occasional enhancement. Photos are often visually striking, specific to a purpose, and contemporary in subject. The major drawback is inadequate size—most are sized for lap or desk viewing. If you need to show a small photo, pass it around; but remember that you will be creating a rolling distraction for some time.

Handouts. For distributed material, the handout is a better alternative than a single photo. Today's copiers make sharp, clear copies. Disruption can be minimized by using an assistant and only a few handout cycles. You may also wish to distribute only

FIGURE 4–13
Show visuals so everyone can see, while you maintain eye contact.

before or after the speech. Such handouts might summarize key points and leave your listeners with data for future reference.

Objects and Models. Although objects and models have dramatic appeal as examples of "the real thing," objects are often too large for portability, scale models too small for visibility. They are effective for smaller groups that can be gathered around for hands-on contact, as with equipment. If you discuss the object's functional makeup, a better alternative might be a schematic diagram or cut-away view.

Demonstrations. You can demonstrate effects better than the mechanisms that produce these effects. Processes, on the other hand, are easier to demonstrate, particularly if they involve human participants. This is the principle used in sports coaching and in the apprentice system in the skilled trades. For a speech, use an assistant rather than a volunteer because of the need for practice.

Slides. Prior preparation and some expense are involved in using slides, but the results can be as good as those produced with fine photographs. Better yet, slides do not disrupt audience attention, particularly if a remote-control device is used to advance the frames.

Films and Videotapes. For the greater expense and the need for more complicated projection equipment, professional-quality results can be obtained. To retain control as a speaker, schedule a discussion and commentary session after the showing. If you need to freeze or replay the action, a tape and a VCR are preferred to a traditional film showing.

Overheads. For flexibility and control, an overhead is one of the best relatively low-tech methods. The projection system allows you to face the audience and to leave the lights on. You can draw on a transparency, overlay it with other transparencies (see Figure 4–14), reveal parts that were initially covered, and shut the projection off instantly. Transparencies are comparatively cheap and easy to produce on standard office copiers.

Computer Screen Projections. An extension of the standard overhead projection may be achieved with a PC-viewer attachment fitted onto a projector linked to a personal computer. Graphic effects produced on the computer can be projected on a screen for the audience. Advances in this kind of multimedia technology are bringing costs down and capabilities up.

Summary

In selecting, designing, and using visuals, follow these guidelines:

1. Use a variety of visuals to achieve these key purposes:

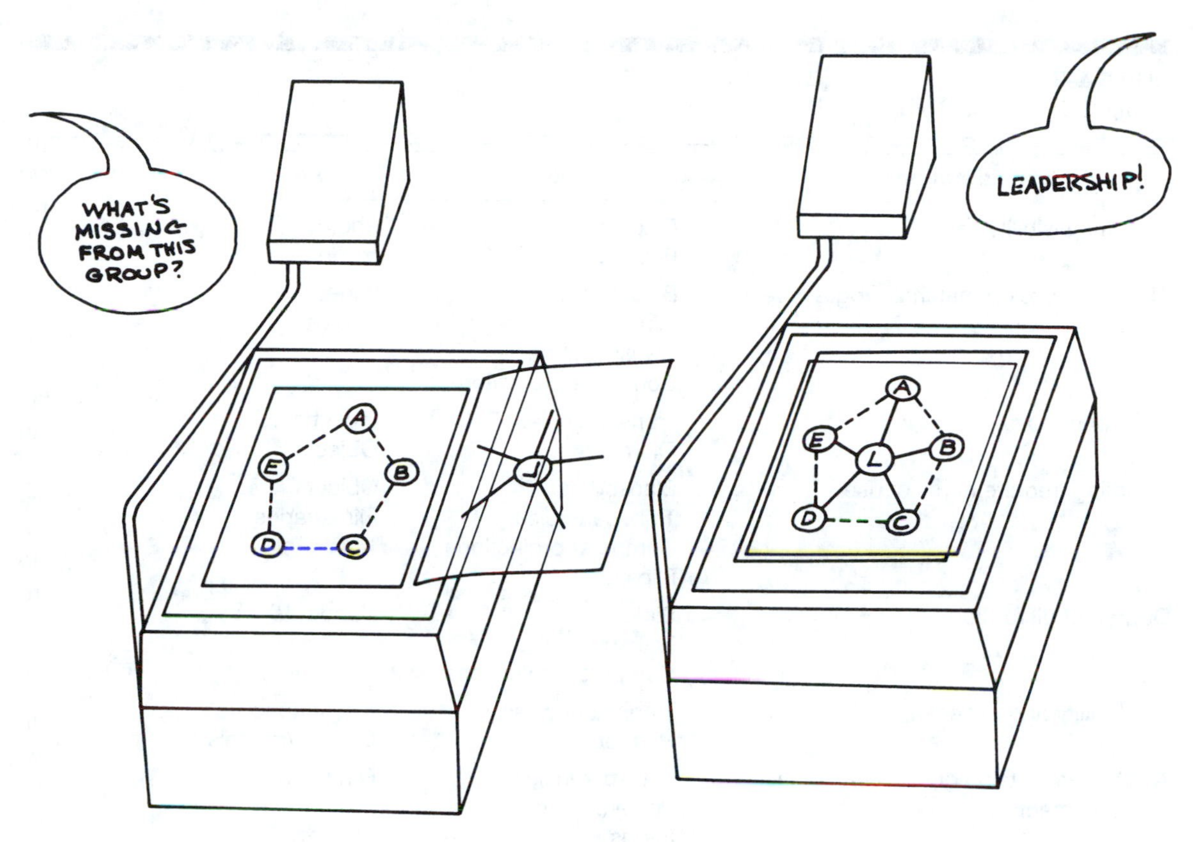

FIGURE 4–14
Using the overlay technique with an overhead projector.

   a. Clarify
   b. Break up
   c. Liven up
   d. Emphasize
2. Plan where to introduce visuals and what you will say about them.
3. Make sure the audience can see the visual; maintain eye contact with the audience; put the visual away when you're through with it.
4. Design visuals to achieve the following characteristics:
   a. Large scale
   b. Limited to one idea
   c. Clear and readable at a glance
   d. Neat and precise
   e. Visually interesting
5. Select visuals according to their strengths and limitations (see Table 4–2).

TABLE 4–2
Strengths and Limitations of Visuals.

Considerations	Good	Poor
Size and visibility	Blackboard Posters	Photos Models
Ability to develop material progressively	Blackboard Flip chart Transparencies Computer projections	Slides Photos
Ease of removal	Transparencies Flip chart	Blackboard Objects
Handling, requirements of use	Blackboard Transparencies Computer projections Flip chart	Objects Films/tapes
Degree of disruption	Slides Transparencies Computer projections	Handouts
Availability and expense	Transparencies Handouts	Films/tapes Objects/models
Ability to maintain control: eye contact and commentary	Transparencies Posters Slides	Films/tapes
Ability to achieve striking visual effects	Films/tapes Photos/slides Posters Computer projections	Blackboard Flip chart

SECTION 5

Research and Development

The development process and development methods discussed in the Introduction and in Section 1 can be enhanced through the specialized techniques presented in this section.

Research techniques, outlining, and document design strategies are planning tools that support development of your ideas through the discipline they impose. The units on plagiarism and proper citation of sources include considerations of form as well as the need for a clear grasp of the material you develop. Word processing techniques are simply tools for enhancing the process of thinking about and reshaping your material during development.

These units are available for practical information as you plan and develop reports or other projects. They can also help you better understand the material you are developing and help you present it more effectively.

UNIT 5.1 RESEARCH

The key to successful projects and research reports is to make use of the wealth of available information and ideas to help you get started and to provide support along the way. In this age of information, for almost any subject you address, work has already been done and the results published somewhere. Gaining access to these results can help you identify new solutions for problems that are not new. Your task to a large extent becomes one of finding the information that can help make your job easier.

In school, the research process is facilitated by a library designed to support the subjects taught in the curriculum. The school library is also linked to a wider network of sources through information retrieval systems and loan arrangements with larger libraries. Electronic information systems have expanded dramatically in recent years. Many libraries now offer access to a variety of data bases, such as ERIC (Educational Resources Information Center); on-line data banks such as DIALOG; CD-ROM (Compact Disk Read-Only Memory) systems that index abstracts and full-text/full-image journal articles; and even global networks such as the Internet and the World Wide Web.

Specialized libraries, sometimes located within the company itself, are also available to people in industry.

A visit to the library, however, is not the first step in the research process, either at college or in industry.

Focusing on the Subject

The first step in academic research is to decide what you want to do. This means selecting a subject to investigate and write about and then limiting that subject to a scope you can handle. In the research-and-development cycle in industry, where you are presented with a general description of a problem to solve, the first step is to define the expected solution more precisely.

In both cases, then, the first research step you should take is to question the "ultimate user" of your work. Many instructors build this step into the development process by requiring students to submit a brief description of their intended project. If your instructor doesn't do so, take the affirmative step yourself. Discuss your idea with her and ask for suggestions. To help focus your thinking, talk to classmates or co-workers as well.

The Literature Search

After you have a more specific idea of what you want to do, you are ready to conduct a literature search. In school, this means a visit to the library, and your first step there should be to ask a staff member where the library's indexes to published literature are (see Figure 5–1). Most of these indexes list periodical articles rather than books

- ❑ Indexes to published literature: article citations and abstracts
- ❑ Business indexes: company and product information
- ❑ Data manuals: schematics, technical specifications
- ❑ Library card catalog or computerized catalog: for books
- ❑ *Books in Print:* for recent books
- ❑ On-line data banks: e.g., DIALOG
- ❑ CD-ROM systems: printouts of article citations, abstracts, full-text articles
- ❑ Internet and World Wide Web resources

FIGURE 5–1
Resources for a literature search.

and will therefore give you an overview of the most current information. This emphasis on periodicals is especially helpful if your subject falls into an area of rapid technological change and of changing industry practices.

Although holdings may vary, college libraries that support business and technical programs offer most of the following indexes, most of which arrange items by subject:

- ☐ *Applied Science and Technology Index* (articles only)
- ☐ *Association for Computing Machinery (ACM) Guide to Computing Literature* (books and articles)
- ☐ *The Business Index* (indexes, books, and articles)
- ☐ *ABI/Inform* (business article abstracts, full-text articles)
- ☐ *Business Periodicals Index* (business articles and company descriptions)
- ☐ *Business Publications Index of Abstracts* (abstracts of business articles)
- ☐ *Computer Literature Index* (abstracts of books and articles)
- ☐ *Engineering Index* (abstracts of books and articles)
- ☐ *Humanities Index* (articles only)
- ☐ *Magazine Index* (general articles, abstracts)
- ☐ *Magazine Express* (abstracts and general articles)
- ☐ *New York Times Index* (newspaper articles)
- ☐ *Readers' Guide to Periodical Literature* (general-interest magazines)

For students researching career opportunities, college libraries and graduate placement offices carry the following kinds of business indexes, which furnish information on companies and industries:

- ☐ *Career Opportunity Index* (regional summaries of employers in technical fields)
- ☐ *MacRae's* state industrial directories
- ☐ *Moody's Industrial Manual* (company descriptions)
- ☐ *Peterson's Annual Guide to Careers and Employment for Engineers, Computer Scientists and Physical Scientists*
- ☐ *Thomas Register of American Manufacturers* (companies and products)

As you conduct your search for current "literature," keep some note cards handy and write down the bibliographic information for each article and book you want to look into. This will help you request the material from the library and will later help you write bibliography entries and citations.

An alternative resource offered by many libraries is CD-ROM equipment. This technology stores information on compact disks that you can access through personal computers in the library. The ProQuest data base *General Periodicals on Disc–Research I,* for example, indexes and abstracts over 1100 journals and offers full-text/full-image versions of articles from more than 320 journals; the *Business*

Periodicals on Disc–Select Edition indexes and abstracts over 350 titles and offers more than 130 in full-text/full-image. With CD-ROM equipment, you can conduct a search for relevant articles and then receive a computer printout of the citations or the articles themselves. Cited articles can usually can be found on the library's shelves or in microfiche holdings.

After the initial search for information that is largely in periodicals, you should also check the library's catalog, which lists books by subject, author, and title. If you are not familiar with the shelving system for these works, ask a staff member for assistance. Most college and university libraries today use the subject classification system of the Library of Congress to organize their shelves. Letter-number headings such as the following are used:

H	Social Sciences
HB	Economics
HF	Commerce
HF 5001–5376	Business
HF 5601–5689	Accounting

If the library has very few books dealing with your subject, you may wish to consult another index, *Books in Print,* which is a subject guide to many more books than your library is likely to stock. However, you may be able to get such books through an interlibrary loan arrangement or directly from a large public library or a university library in your area.

Once you find articles and books relevant to your subject, you can take notes on them while you're in the library, and you can usually check the books out. Most libraries today offer copying services, and it may be to your advantage to copy articles and relevant portions of books so you can examine the material at more leisure. This will also ensure that you have the full text of material you wish to *quote* in your report.

Specialized Sources

The literature search in industry and for the upper-term design projects you do in school is likely to be more focused from the beginning. You may want to look into current periodicals as well, but your primary attention will be devoted to specialized data manuals that publish manufacturers' specifications and application notes and that provide evaluations and users' ratings of hardware and software. Examples of such manuals follow:

Datapro (on edp systems, micros, communications, and industry and office automation), published by Datapro Research Corporation.

IC Master (on integrated circuits, programmable logic devices, design automation, and microprocessor development systems), published by Hearst Business Communications.

Electronic Engineers Master Catalog (on electronic components, interconnections, power sources, and instrumentation), published by Hearst Business Communications.

For many projects, you can also search a computerized data base, such as DIALOG, for current information in articles and books as well as materials with limited distribution, such as conference proceedings, technical reports, and statistical compilations. In most cases, the search will be conducted by staff in the library subscribing to this service. You will need to analyze your topic and provide key words to focus the search.

Assume, for example, that you are researching the effects of poor organizational communication on employee morale and productivity. If you ask the computer to search for information on topics such as "communication" or "employee morale," you will get very long lists of citations that include much unrelated information. Instead, you can cross two or more topics and thus narrow the search field.

For this project, you might specify "organizational communication" AND "employee morale" AND "productivity" as the search parameters. Use of the AND command will give you only citations that appear in all three categories (see Figure 5–2).

Another kind of research you may have to do during a design project is to determine the regulations that restrict your design. For example, the Federal Communications Commission sets standards for radiation leakage in microwave devices, and the U.S. Underwriters' Laboratories regulate safety standards. In Canada, the Canadian Standards Association regulates certain technical specifications. For such considerations, the data manuals point out applicable rules and list the agencies that oversee these matters. In the design of computerized systems in industry, you also may have to consult your company's legal department for guidance on copyright issues associated with software.

Searching the Internet and the World Wide Web

With access to the Internet, you can also *browse* for information stored on computers around the world. Browsing services use specialized software to locate and

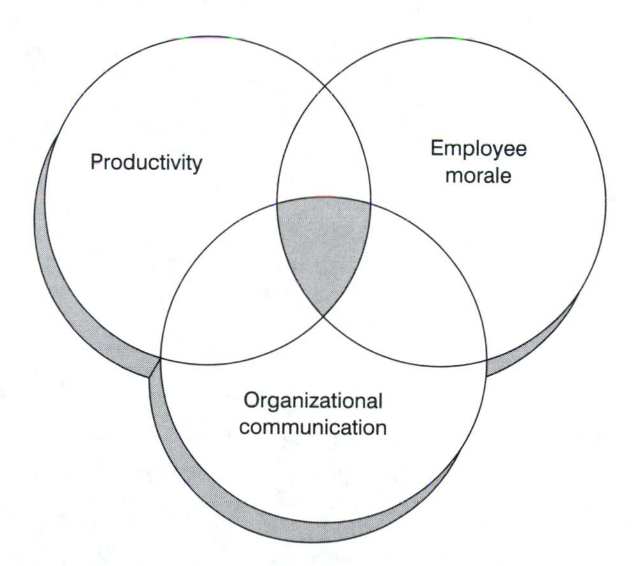

FIGURE 5–2
Computer data base search strategy. Use of three topics linked by the AND command produces only reports that bear on all three topics.

access information stored at remote sites. With a browsing service such as Gopher, you can locate computers in which information is stored, read menus listing various contents, access items you are interested in, and print out items you wish to keep for reference. The range of information accessible by a Gopher search includes college course descriptions, satellite weather maps, economics data, computer programs, lists of books, and scientific information.

Gopher is a menu-driven service. It presents you with lists, or menus, of topics about which information is available. By using the mouse to point to an item and click on it, you retrieve more detailed information, either a more specific menu or, finally, the information itself. Thus, you may "click down" through a sequence such as the following: Top Level Menu—Other Gopher Servers—North America—United States—North Carolina—University of North Carolina at Chapel Hill—Frequently Used Resources. The last screen would then display a menu including the items shown in Figure 5–3.

If you clicked on the *Academe This Week* item, you would access a compilation of news from higher education. At this point you would have reached the final level of information, rather than additional menu items. By clicking on *Previous Menu,* you could travel back toward your starting screen to consider other choices. By clicking on *Bookmarks* at the top of the screen, you could effectively save a Gopher location you wanted to return to later, or periodically, so that you would not have to repeat the search string each time.

With more advanced browsers, such as Netscape, you can search the World Wide Web—the network of Internet computers storing hypermedia documents that combine text, graphics, sound, video, and multimedia combinations of resources. Netscape will let you access pages integrating text and embedded menu titles, with

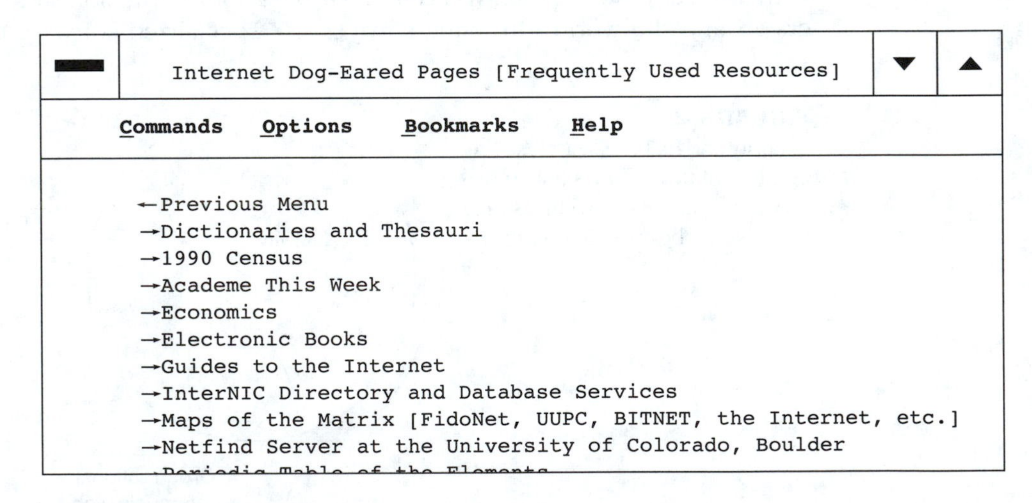

FIGURE 5–3
A computer screen displaying a partial list of menu items from the University of North Carolina's Gopher site on the Internet.

the latter highlighted for further access to more detailed information or even sound or video displays (see Figure 3–16 on page 168).

The Web uses a system of Uniform Resource Locator (URL) codes to identify the location of the stored information. A URL such as http://www.cracker.barrel/, for example, tells Netscape which computer to contact for the desired information. It is like a phone number you can call that helps you avoid the search strings of the Gopher service.

If you don't know the URL code, you can use one of the automated search tools, such as Archie, to search the Internet by file name or by subject or content. If you tell Archie to find all the files named "William Jackson Weiland Jr.'s Excellent Home Page," you may get one listing; if you ask for files with the word *computer* in the title, you may get thousands. As in data base searching, a practical search strategy for Archie is to qualify generic words with limiting terms (e.g., *Internet host computers*).

An automated search service like WAIS (Wide Area Information Server) scans for a particular word or phrase in the texts of documents, not in their titles. Tools like WAIS are helpful because many titles don't reflect their contents in ways that computers can distinguish. While we understand that George Orwell's *1984* is about the threat of totalitarian control, a computer searching for these concepts in titles would not select this work. Nor would a search of the book's contents select the book, since Orwell evokes the concept of totalitarian control figuratively rather than literally. A contents search for the phrase *Big Brother,* on the other hand, would list *1984,* as well as numerous other works influenced by this fictional persona from Orwell's novel.

A Research Philosophy

Sources of information and research tools vary widely, but the basic research process is generally the same in all fields. By learning as much as you can about the process while you're still in college and by developing the willingness to seek information from people, published sources, and electronic information systems and networks, you are developing practical tools that should help you grow professionally throughout your career.

UNIT 5.2 OUTLINING

An outline is an organized summary of the main points of a written or oral presentation. The Introduction section of this book gave you some suggestions on developing a *working outline* to guide your writing of the first draft. The working outline can serve as a script for the draft you wish to produce.

This unit explains the features of the *formal* outline. Your instructor may ask you to submit a formal outline to check your progress on a report assignment. A detailed formal outline can also serve as a basis for the table of contents in formal reports, user's manuals, and other long projects.

A grasp of outlining principles will help you strengthen the organization of your reports. The outline not only keeps you on track but also shows you where additional development is needed. It can even expose logical flaws.

Subordination and Parallelism

An outline divides and subdivides a subject into its major parts. The parts then relate to each other (and to the general subject) in two basic ways:

1. Some parts are parallel in importance to other parts.
2. Some parts are the elaborations or subdivisions of other parts.

Consider the following example:

A. Consumer products
  1. Convenience goods
  2. Shopping goods
  3. Specialty goods
B. Industrial products

This cluster would be somewhere in the middle of the outline. Main points A and B are elaborations or divisions of some superior part headed by a Roman numeral (e.g., II. Marketing strategy). Point B is parallel in importance to point A. Subpoints 1, 2, and 3 are elaborations or divisions of point A. Subpoints 1, 2, and 3 are also parallel in importance to one another.

There must always be at least *two* subparts to any superior part. The use of only one subpart suggests that the division hasn't been completed or that it can't be done. A single subpart should be either eliminated or incorporated into its superior part. Thus:

B. Social sciences
  1. Psychology
	should be revised to	B. Social sciences: psychology
		or
		B. Social sciences
		  1. Psychology
		  2. Sociology

The concepts of subordination and parallelism can be helpful planning tools even before they are applied to outlines. If I am thinking about a report I have to write on fields of study, for example, I might generate a random list of subjects such as the following:

anthropology	sociology	abnormal psychology
biology	zoology	group therapy
physics	literature	art history
chemistry	electronics	painting
psychology	psychoanalysis	sculpture

economics
philosophy
art
social psychology
family therapy
encounter groups

Such a list obviously needs some organization, which will also help me focus my plan for the report. I might begin to group the topics as follows:

anthropology
psychology
economics
sociology
social psychology
psychoanalysis
group therapy
abnormal psychology
family therapy
encounter groups

zoology
biology
physics
chemistry
electronics

literature
philosophy
art
art history
painting
sculpture

My second grouping indicates that I have assembled topics in three primary categories that I might call social sciences, natural sciences, and humanities. But how does "group therapy" relate to an item such as "anthropology"? Is it parallel or subordinate? As I sort through my topics, I begin to see that I have assembled different categories of items, at different levels. My *outline* shows me the proper relationships:

A. Social sciences
  1. Anthropology
  2. Sociology
  3. Economics
  4. Psychology
    a. Social psychology
    b. Abnormal psychology: therapies
      (1) Psychoanalysis
      (2) Group therapy
        (a) Family therapy
        (b) Encounter groups
B. Natural sciences
  1. Zoology
  2. Biology
  3. Physics: electronics
  4. Chemistry
C. Humanities
  1. Literature
  2. Philosophy

3. Art
a. Art history
b. Painting
c. Sculpture

The outline also shows me that I have much more detail (i.e., subordination) in some areas than others. If I want to present a balanced report on all areas, I need to develop subordinate items for many of the other fields. Conversely, I might write a report that briefly touches on the general categories, but focuses in depth on only one, such as psychology.

Outline Form

Two common types are the topic outline, which consists of phrases, and the sentence outline, which uses complete sentences. Use either all sentences or all phrases in an outline. This consistency of form, or parallelism, supports the equality of content between parallel items in an outline.

In both topic and sentence outlines, start entries with a capital letter. End sentences with a period and phrases with no punctuation. If a sentence or phrase runs longer than one line, start the second line directly below the beginning capital letter of the first line:

I. Sentences or phrases in an outline can be longer
than one line.

Place subdivision indicators (A., 1., a., etc.) directly below the beginning capital letter of the superior sentence or phrase:

I. Superior sentence or phrase
A.

The most common outline notation uses a sequence of letters and numbers according to the scheme illustrated in a middle section of the following topic outline:

II. Distribution channels
A. Retailers
1. Types of retailers
a. Department stores
(1) Product variety
(a) Clothing
i) Children's
ii) Men's
iii) Women's
(b) Housewares
(c) Hardware
(2) Service

(3) Value: price/quality
b. Specialty stores
c. Convenience stores
d. Discount stores
2. Retail strategies
B. Wholesalers
III. Promotional strategy

Notice that each subdivision has at least two parts. Note also the way the Roman numerals align vertically on the decimal point (or parenthesis sign) instead of the first digit.

A popular alternative outline format uses the decimal notation method, which works as follows:

1.0 Retailers
1.1 Types of retailers
1.1.1 Department stores
1.1.1.1 Product mix
1.1.1.2 Service
1.1.1.3 Value
1.1.2 Specialty stores
1.1.3 Convenience stores
1.1.4 Discount stores
1.2 Retail strategies
2.0 Wholesalers

An advantage of decimal notation is that each item becomes easy to refer to because its notation includes the notations of all previous items of higher rank. A disadvantage is that the numbering sequence becomes cumbersome after more than three or four levels (e.g., 1.2.4.1.3).

Decimal notation is also used in many textbooks, especially those dealing with technical and scientific subjects. In these cases, the digits before the first decimal point refer to the chapter number, and the digits after the decimal point indicate the section of the chapter. Subsections can also be indicated with this method. (The units in this text are labeled according to decimal notation.)

In technical textbooks with many equations, decimal notation can also be used for numbering the formulas consecutively in parentheses at the right margin:

$$R_s R_p = \frac{L}{C} \qquad (13.9)$$

In this case, the notation indicates the ninth main formula in Chapter 13.

Decimal notation in textbooks makes reference to chapters, chapter sections, and formulas both easier and faster.

UNIT 5.3 USING SOURCES

Plagiarism is the use of someone else's ideas, information, words, or other means of expression without giving credit to the originator. Without acknowledging a source, you create the impression that the material is your own. That, in plain terms, is theft.

Obvious, and extreme, cases of plagiarism occur when a journal article or an encyclopedia entry is copied word-for-word and submitted as one's own work. Plagiarism also occurs when a report is prepared by someone other than the listed author—by another student, a co-worker, or a term paper "service."

In most cases, plagiarism occurs on a more limited scale; that is, parts of someone else's work are incorporated into a report without citing a source. In some cases the borrowing is unintentional.

It's fair to say that most writers are interested in avoiding plagiarism, but may not always be clear on the difference between plagiarism and legitimate use.

In determining what belongs to someone else, keep in mind that many ideas and much information are *common knowledge.* The basic laws of physics and electronics, for example, are in the public domain, even though particular individuals originated these ideas at one time. You do not have to cite a source for ideas or information of this kind, even though you may have had to look it up yourself. The basics of a discipline or career field are also "common knowledge" even though many people outside the field are unfamiliar with them.

On the other hand, you must give credit for the following kinds of material outside the domain of common knowledge:

1. Someone else's *ideas*—a deduction, inference, or generalization from known or observed data; a design; a judgment ("This is the most efficient device/the best value. . . . ")
2. Specific *facts* or *data*—research results such as those from tests, measurements, observations, and surveys
3. Someone's *expression* of facts or ideas—specific wording, a particular graphic design, or a unique presentation of both words and graphics.

Using Sources Properly

If you want to use someone else's ideas or information, cite your source through a reference (see Unit 5.4, Bibliographies and Citations). If you want to use someone else's exact words, put quotation marks around them and refer the reader to the source. If you want to use someone's graphic design, you can copy it or redraw it if you include a note citing the source (see Figure 5–4). But if your work is being submitted for publication or is for commercial purposes, you will need to obtain written permission from the copyright holder to use the material. If you want to use someone's information or ideas but not his words, you can *paraphrase* the original and then cite its source.

To paraphrase is to restate something in your own words. To make sure the words are really your own, try to absorb the key points in the passage and then restate them in a way that fits your normal style. It is not enough to just change a word or a phrase here and there.

FIGURE 5–4
Credit must be given (and permission obtained) for trademarks, which are unique combinations of words and graphics. Source: Hartford Fire Insurance Company. Reprinted with permission.

The following examples show proper and improper forms of paraphrasing:

Original Passage

Just-in-time training differs from a traditional classroom approach, in which participants learn concepts and skills and later have to figure out how to apply them back on the job. For one thing, a traditional approach to quality-improvement training requires training employees first, before they can participate in an improvement project. That extra step means that the learning process requires a lot more time, perhaps delaying improvement results.[1]

Improper Paraphrase

Just-in-time training is different from traditional approaches, in which classroom participants learn skills and concepts they later have to figure out how to apply back on the job. For example, the traditional methods of quality-improvement training call for training employees first, and then letting them take part in improvement projects. That creates an extra step and takes a lot more time, and may delay improvement outcomes.

The preceding version fails as a paraphrase because it retains much of the original wording and structure with only minor changes in a few places.

Proper Paraphrase

Just-in-time training offers significant advantages over traditional methods of quality-improvement training. The traditional approach creates an extra step, and possible delays, by taking employees away from their jobs and putting them into a classroom. This learning must later be transferred back to the workplace.[1]

---

1. Richard Y. Chang, "When TQM Goes Nowhere," *Training & Development* 47 (January 1992): 26.

In this version, the original passage has been sufficiently rewritten to become a valid paraphrase. Notice that a citation for the passage is still required because the paraphrase contains inferences and judgments about techniques that are not common knowledge and that come from a particular source.

Unit 5.3 Exercise

1. Which of the following restatements represents a proper paraphrase of the quoted lines?

   In determining what belongs to someone else, keep in mind that many ideas and much information are common knowledge. *The basic laws of physics and electronics, for example, are in the public domain, even though particular individuals originated these ideas at one time. You do not have to cite a source for ideas or information of this kind, even though you may have had to look it up yourself.*

   a. To decide what belongs to others, remember that many ideas and facts are in the category of common knowledge. For example, the fundamental laws of physics and electronics are in the public domain, even though certain persons discovered these laws once upon a time. You do not have to cite sources for ideas like these, even though you may have had to look them up.

   b. Ideas and information that are commonly available to the public do not have to be credited to a source, even if you had to look these ideas up yourself. Such "public domain" material includes basic physics and electronics principles that, admittedly, were conceived by specific persons.

2. Does the properly paraphrased version of the original lines have to be referenced to a source? Why or why not?
3. Try writing your own paraphrase of the original lines above. Then compare your version to the original and to the paraphrase you selected in Question 1. Are you saying the same things as they are, except in your own words?

UNIT 5.4 BIBLIOGRAPHIES AND CITATIONS

Formal reports, proposals, feasibility reports, and other major writing projects may require that you cite (or credit) the sources of specific facts, figures, or opinions as they occur in your report. The same projects may also require that you provide a list of all the books, manuals, articles, and other sources you consulted in preparing the report—whether you used specific information from them or not. This list is called a bibliography.

Citations: The Method

Several widely used style manuals provide guidelines on all aspects of manuscript preparation, including citations and bibliographies. Broad agreement on matters of form—along with some variation—exists among such guides as the *Publication Manual of the American Psychological Association, MLA Handbook for Writers of Research Papers* (of the Modern Language Association), and *The Chicago Manual of Style* (University of Chicago Press). While this book follows the guidelines of *The Chicago Manual of Style,* there is little disagreement among these manuals on how to cite sources properly.

The preferred method of citation today is to enclose a brief reference to the source in parentheses within the sentence that gives the information. In this method, the reference is also linked to a bibliography at the back of the report, where complete information about all sources is given. In the parentheses, give only the author's last name, the year of publication, and, if needed, the page number:

> Some viruses are designed to remain dormant in a system until a particular program is run a certain number of times (Silver and Silver 1991, 239).

This parenthetical reference can be checked in the bibliography, where the following book will be listed: Silver, Gerald H., and Myrna L. Silver. *Data Communications for Business.* 2nd ed. Boston: Boyd & Fraser, 1991.

If the author's name is not available, a key word or words from the title may be used instead: (*Data Communications,* 239). Because some bibliographies may contain several books or articles by the same author published in the same year, you may need to add a letter to the year of publication to help distinguish the work:

> Silver and Silver 1991a, 239

While the parenthetical citation method is most widely used, you should also be familiar with some of the older methods that are still used in some organizations, and may be required of you as well. In the traditional citation method, a reference to a source is indicated by a raised number (or superscript) in the text immediately after the information being cited. These superscripts are numbered in sequence through the entire report. Each superscript is referenced to a listing that includes the author, title, publishing information, and page number of the source.

For these listings, you may prefer to enter *footnotes* at the bottom of the page, or you may provide a list of *endnotes* after the conclusion of your report. The ending list is then titled "Notes" or "References." A model of this method for the twelfth citation in a report follows:

TEXT INDICATION	One of the heaviest commercial users of audio conferencing may be Intel Corporation, which has 600 specially equipped conference rooms and puts through about 7,000 conference calls each month.[12]
CITATION	12. Elliot M Gold, "Without the Audio, There Is No Conference," *Networking Management* (December 1992): 47.

Bibliography: The Method

In listing all consulted sources, whether cited or not, you are preparing a bibliography. Arrange your sources alphabetically by the last name of the author (or editor). For unsigned articles, use the first major word of the title to alphabetize.

You will find the bibliographic information about a work on the title page at the front of the book and on the back of the title page, where the year of publication is listed. For a bibliography, divide the author, title, and publishing information by periods (not commas) according to the following model:

> Pfeiffer, William S. *Proposal Writing: The Art of Friendly Persuasion.* Columbus, OH: Merrill Publishing, 1989.

Other varieties of sources are shown in the bibliography form section, in Figure 5–5.

Bibliography Form

Figure 5–5 shows examples of several kinds of sources as they would be listed in a bibliography prepared according to each of the three major styles: (a) *The Chicago Manual of Style* of the University of Chicago Press, (b) MLA (Modern Language Association), and (c) APA (American Psychological Association).

Citation Form

If each of the sources listed in the bibliographies in Figure 5–5 had been cited once and in the same order, and all citations had been listed at the back of the report, the result would be a "notes" page like the one in Figure 5–6.

UNIT 5.5 WORD PROCESSING AND DOCUMENT DESIGN

Writing is a process of planning, development, revision . . . and redevelopment . . . and re-revision. . . . In this situation, word processing offers the ultimate tool and relief agent. It not only gives you endless flexibility in revision but also allows you to format the text variously, add graphics, and print your writing in various type styles. This flexibility and ease of revision allows you to concentrate on the substance of your message so you can make it as clear and effective as possible. Because word processing is also *fun,* it can increase your enjoyment of writing itself.

This unit briefly explains the major capabilities of word processing and suggests some guidelines for using this technology to design more readable documents.

Word Processor Versus Personal Computer Versus Desktop Publisher

You can process your writing on either a dedicated word processor or a personal computer. In the latter case, your computer would read a disk containing a word processing program (see Figure 5–7). With other kinds of disks, the same computer would perform other functions, such as budgeting, planning, or record keeping.

A dedicated word processor, on the other hand, is a computer that has already been programmed to do word processing via instructions built into its central processing unit.

The highest level of word and text processing capability today can be achieved through a desktop publishing system consisting of a powerful microcomputer and

	A. *Chicago Manual of Style, University of Chicago Press*
Book: One Author	Comer, Douglas E. *The Internet Book*. Englewood Cliffs, NJ: Prentice-Hall, 1995.
Book: Several Authors, Later Edition	Courtney, James F., Jr., and David B. Paradice. *Database Systems for Management*. 2nd ed. Boston: Irwin, 1992.
Book: With Editors	Burke, John G., and Marshall C. Eakin, eds. *Technology and Change*. San Francisco: Boyd & Fraser, 1979.
Article: With Author, Journal	Genet, Russell M. "Updating the Oldest Science." *BYTE* 10 (July 1985): 179–191.
Article: Unsigned, Popular Magazine	"Multimedia Authoring Products." *Syllabus,* November 1995, 48–53.
Encyclopedia Article	Not included in bibliography but cited in text: "According to the entry on microsurgery in the *New Encyclopaedia Britannica,* 15th edition, micromanipulative techniques were . . ."
User's Manual	Hewlett-Packard Company. *HP LaserJet 4 and 4M Printers User's Manual*, October 1992.
Electronic Information Service	Breland, Hunter M., *Assessing Writing Skill* (New York: College Entrance Examination Board, 1987), ERIC ED 286920.
Interview	Fermi, Enrico. Telephone interview by author. 29 August 1958.
	B. *MLA Handbook for Writers of Research Papers,* Modern Language Association (MLA style)
Book: One Author	Comer, Douglas E. *The Internet Book.* Englewood Cliffs: Prentice, 1995.
Book: Several Authors, Later Edition	Courtney, Jane, F., Jr., and David B. Paradice. *Database Systems for Management*. 2nd ed. Boston: Irwin, 1992.
Book: With Editors	Burke, John G., and Marshall C. Eakin, eds. *Technology and Change*. San Francisco: Boyd & Fraser, 1979.
Article: With Author, Journal	Genet, Russell M. "Updating the Oldest Science." *BYTE* 10 (1985): 179–91.

FIGURE 5–5
Examples of entries in a bibliography according to each of the three major styles.

Article: Unsigned, Popular Magazine	"Multimedia Authoring Products." *Syllabus* Nov. 1995: 48–53.
Encyclopedia Article	"Microsurgery." *The New Encyclopaedia Britannica: Micropaedia.* 15th ed. 1991.
User's Manual	Hewlett-Packard Company. *HP LaserJet 4 and 4M Printers User's Manual*. 1992.
Electronic Information Service	Breland, Hunter M. *Assessing Writing Skill.* New York: College Entrance Examination Board, 1987. Online. ERIC ED 286 920. 22 Feb. 1996 [i.e., the date you accessed the material online].
Interview	Fermi, Enrico. Telephone interview. 29 August 1958.
	C. *Publication Manual of the American Psychological Association* (APA style)
Book: One Author	Comer, D. E. (1995). *The Internet book*. Englewood Cliffs, NJ: Prentice-Hall.
Book: Several Authors, Later Edition	Courtney, J. F., Jr., & Paradice, D. B. (1992). *Database systems for management* (2nd ed.). Boston: Irwin.
Book: With Editors	Burke, J. G., & Eakin, M. C. (Eds.). (1979). *Technology and change*. San Francisco: Boyd & Fraser.
Article: With Author, Journal	Genet, R. M. (1985). Updating the oldest science. *BYTE, 10,* 179–191.
Article: Unsigned Popular Magazine	Multimedia authoring products. (1995, November). *Syllabus,* pp. 48–53.
Encyclopedia Article	Microsurgery. (1991). In *The New Encyclopaedia Britannica* (Vol. 8, p. 105). Chicago: Encyclopaedia Britannica.
User's Manual	Hewlett-Packard Company. (1992) *HP LaserJet 4 and 4M printers user's manual.*
Electronic Information Service	Breland, H. M. *Assessing writing skill* [Online]. ERIC ED 286 920.
Interview	Not included in bibliography but cited in text: "Enrico Fermi (personal communication August 29, 1958) suggested this to the author."

FIGURE 5–5, *continued*

Chicago Manual of Style style:

NOTES

1. Douglas E. Comer, *The Internet Book* (Englewood Cliffs, NJ: Prentice-Hall, 1995), 152.
2. James F. Courtney, Jr., and David Paradice, *Database Systems for Management*, 2d ed. (Boston: Irwin, 1992), 376–77.
3. John G. Burke and Marshall C. Eakin, eds., *Technology and Change* (San Francisco: Boyd & Fraser, 1979), 233.
4. Russell M. Genet, "Updating the Oldest Science," *BYTE* 10 (July 1985): 187.
5. "Multimedia Authoring Products," *Syllabus*, November 1995, 48–53.
6. "Microsurgery." *Encyclopaedia Britannica*, 15th ed., s. v.
7. Hewlett-Packard Company, *HP LaserJet 4 and 4M Printers User's Manual,* October 1992.
8. Hunter M. Breland, *Assessing Writing Skill* (New York: College Entrance Examination Board, 1987) ERIC ED 286920.
9. Enrico Fermi, telephone interview by author, 29 August 1958.

FIGURE 5–6
Sample entries in a list of notes.

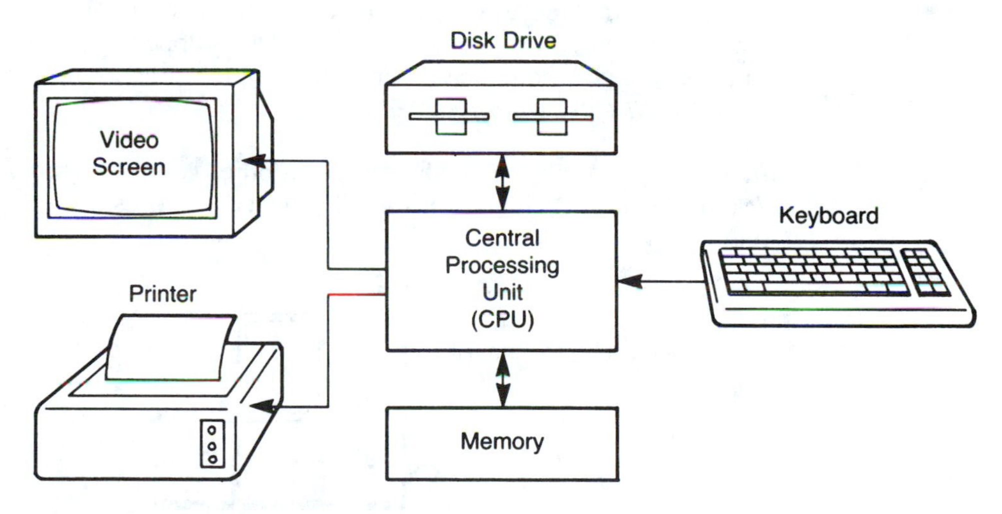

FIGURE 5–7
The parts of a word processing system.

specialized software, in addition to attachments such as a graphics tablet or mouse. The desktop publishing system offers not only word processing but also greater graphics capability, a wide range of type styles and sizes, and a full integration of text and graphics. Organizations using such systems can now produce visually impressive reports as well as internal publications, manuals, and marketing brochures. The results are virtually indistinguishable from those offered by professional typesetters.

If current trends in software development continue, the capabilities of desktop publishing systems will be increasingly available in standard word processing packages.

Word Processor Capabilities

Word processors and word processing packages may have more or fewer capabilities and almost certainly have different ways of executing their capabilities via the keyboard. Certain basic operations, however, are available in nearly all systems.

Text Entry

1. You type characters on the keyboard as you would on a typewriter. What you type appears on the screen.

2. A *cursor* (a flashing symbol) moves across the video screen, indicating where you are at the moment. By using cursor movement keys (see Figure 5–8), you can shift the cursor right or left and up or down to a new location where you may want to make changes.

3. A *word wrap* feature lets you type without concern for the length of a line. A word returns to the beginning of the next line after the cursor hits a preset position at the right of the page.

4. *Scrolling* moves your text up or down the screen for viewing. Your document "unrolls" in either direction. As earlier lines appear at the top, for example, later lines disappear from the bottom. You can also "scroll" a whole page at a time by using the *PgUp* and *PgDn* keys.

Editing. By moving the cursor and using other control keys, you can *insert* or *delete* material of any length. Surrounding material "moves over" for insertions and "contracts" for deletions.

FIGURE 5–8
Cursor movement keys for repositioning the cursor.

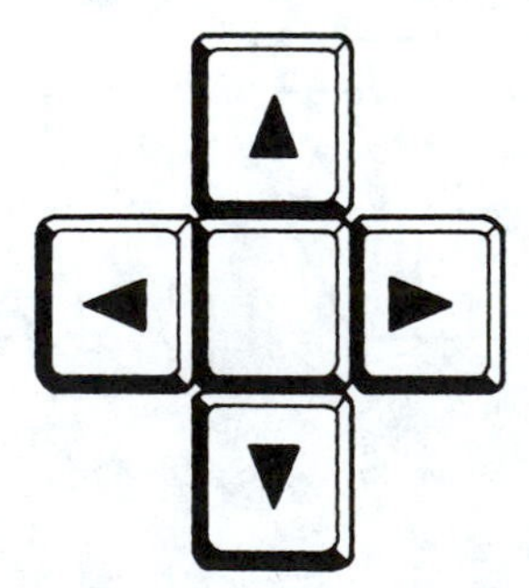

You can also *move* parts of your document to other locations, or even to other documents. In the days before word processing, this was known as "cut and paste." Today you can quickly move a sentence from one part of a paragraph to another, or move a whole paragraph or block of paragraphs. The result should be better organization and a smoother flow of your words.

You can *search* for a sequence of characters, a word, or a phrase anywhere in the document, and *replace* it with other material. Search and replace is a handy feature for misspelled words, misspelled names of people or companies, and errors in figures such as measurements or in calculated values.

A variant of this feature is the *spelling checker,* which may be part of a word processor or may be added via a package. Although most checkers do not correct the spelling for you, they do identify questionable words. Spell checkers come with dictionaries of up to 50,000 words, and some allow you to add your own words to the list.

Formatting the Text. You can *justify* your lines, that is, align them vertically, both at the left and the right of the page. (Right justification, however, should be used with caution since it may produce large and irregular gaps between words. This may happen if you are using a short line length, as in the case of a two-column page (see Figure 5–9).

You can *center* material across the page. This is useful for titles and for equations or phrases set off from the text:

Work and Energy

. . . and this effect is proportional to the square of the velocity, as shown in Leibnitz's formula:

$$F\Delta s = \frac{mv^2}{2}$$

You can change *line spacing.* Let's say you want to produce a double-spaced version of your report so a supervisor can make changes in it, but the final version has to be single spaced. You can tell the computer to print the report either way.

Printing the Text. Error-free documents are possible because all corrections can be made on the screen before printing. At the least, the mistakes will not come from the printing process, as they do in typing.

FIGURE 5–9
A two-column format with right justification. The short lines and lack of proportional spacing produce uneven results.

```
You can justify lines,     should  be  used  with
that  is,  align  them     caution   because   it
vertically    both   at     may   produce   large
the left and the right     and   irregular   gaps
of   the   page.  Right    between         words.
justification, however,
```

You can print all or selected parts of a document, and you can combine several documents into one.

Special effects are possible. You can get **boldface,** *italics,* different **type styles and Sizes,** superscripts (y^3), subscripts (A_1), and mathematical symbols (μ, Σ). With advanced wordprocessing packages you can also print text with *proportional spacing,* which is the way this book is typeset. Notice that the *i* in *time* gets less space than the *m.* Proportionally spaced text looks better and reads easier.

Graphics. Some of the most impressive capabilities of modern word processors are found in graphics production and the integration of graphics with text. You can create charts on the video screen by controlling the *pixels,* or minute dots, that form all characters or lines. You can size a chart to fit, label axes, and make all the adjustments you need before printing it out together with your text. Two common types of graphics programs are available:

1. *Analysis graphics,* as the name implies, analyzes numerical data you furnish and then puts the data into visual form by creating bar charts, pie charts, line graphs, scatter diagrams, and other standard forms.

2. A *graphics editor,* or *paint package,* gives you additional design capability in creating even more sophisticated visuals. You are the designer because you select and assemble a variety of predrawn figures from the computer's memory to produce a customized graphic. You can even create freehand pictures or "paint" directly on the screen by using a stylus and graphics pad. In the chart shown in Figure 5–10, the visual elements available from memory include the shading for the rectangles and circles. Note also the diagonal lines drawn with the graphics pad, and the freehand signature of the designer.

Word Processing and Document Design

As the preceding discussion indicates, word processing gives you options in the formatting of text, in type styles and sizes, and in graphics capabilities. All of these options can be applied to the task of *designing* the letters, memos, and reports you compose.

In the design process, you need to make some general decisions about your document—margins, line spacing, and type style and size, for example. Specific attention may then be devoted to the layout of material on each page to enhance its readability. Document design, like the design of buildings or cars or clothes, should follow the principle that form serves function.

One of the most useful strategies for effective layout of a page is to divide it into blocks to be filled by three kinds of elements: words, visual elements, or just white space. The last element is, in design terms, one of the most important. White space is the necessary medium for *presenting* words and graphics. Unless white space surrounds them, words and graphics become illegible.

A basic block of material to consider is the paragraph. You should change paragraph length for the sake of variety, as you do the length of sentences. On the other

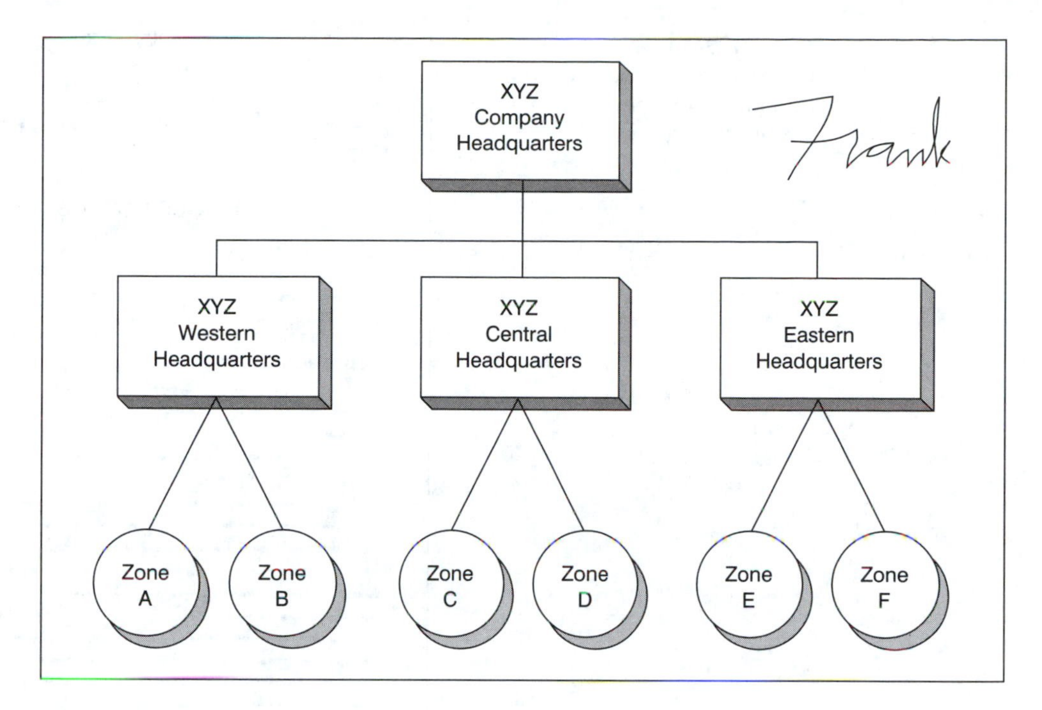

FIGURE 5–10
An organizational chart created on a desktop publishing system using a paint package. Source: Amgraf Inc. Reprinted with permission.

hand, try to keep paragraphs five to six sentences long on average so you can provide frequent breaks for the reader by either indenting first lines (as this text does), separating paragraphs with a line of white space, or both.

Larger blocks of the material may include a group of paragraphs that make up a section of your text. For these you can select a variety of subheadings in a more prominent type style and a larger size, and combine them with white space to divide the material. Generally, for major sections, use headings that are centered across the page; for subsections, use subheads that start at the left margin. Use all capitals for centered headings, upper- and lowercase for subheads. Underline subheads or use italic or bold type (see Figure 5–11).

To highlight material, set it off from the rest of the text. Bullet-pointed or numbered lists of parallel items are a common example of a technique that not only breaks up the text but presents suitable material in more readable form. Quotations longer than a few lines may also be set off from the text, in a block format with all lines indented (see Figure 5–12).

Figures, tables, and other graphics are a third major component that can be mixed with text and white space to help you balance the page. While basic principles for use of graphics are presented in Unit 1.8, word processing gives you additional design options. By setting variable line lengths for your text, you can also present

FIGURE 5–11
Coordinated use of paragraphs, headings, and white space in the layout of a page.

6

MAJOR HEADING

Subheading

graphics and text side by side across the page, creating the effect of wrapping text around your visuals. Variants of this technique, including a two-column layout, are shown in Figure 5–13.

A good model for document design strategies is the front page of a newspaper. Newspapers use headlines, text, and visuals to achieve three main purposes:

- Feature some stories and story elements
- Guide the reader's eye around the page
- Provide variety and breaks for the reader

With the design options provided by word processing technology, you can use text, graphics, and white space to achieve similar results in your documents.

FIGURE 5–12
A page layout including a bullet-pointed list and a quotation set off from the text.

7

– Source/Author

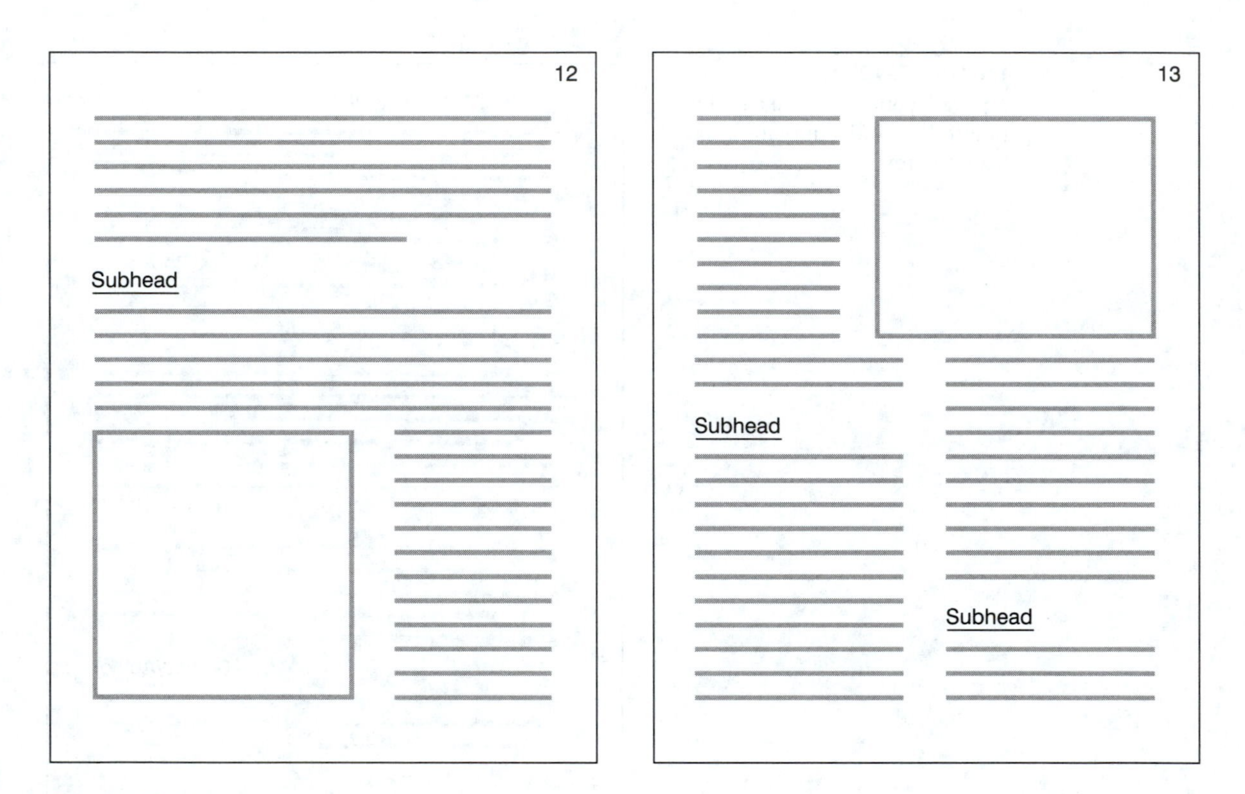

FIGURE 5–13
Two approaches to combining text with graphics, subheads, and white space.

SECTION 6

Grammar, Usage, and Mechanics

Use the material in Section 6 for reference when you revise. The topics included here should supplement your own knowledge and your ear for grammatical language. Most of us use expressions that "sound right" and avoid those that don't. We say "he doesn't" instead of "he don't" without conscious reference to grammatical rules (unless we grew up listening to Uncle Max, who always said "he don't"). On the other hand, we may have to think a moment before choosing "between you and me" instead of "between you and I." If you have to think too long about too many of these rules, study the examples and then try the exercises at the end of Units 6.1, 6.2, 6.3, and 6.4.

The best way to strengthen your command of basic grammar and usage is to write first and then adjust your efforts. Having something to say provides an incentive for learning the grammatical forms that will minimize interference with your message.

Five of the units in this section—Units 6.3 through 6.7—deal with common questions of form in writing. Most of these questions are covered by rules or at least well-established guidelines. Spelling is perhaps least served by rules and could use them most. But even spelling offers specific answers to the questions of form. Most words have one spelling, even if it is not always logical.

The rule-bound nature of these form matters makes them "mechanical." Yet, mechanics can also affect comprehension when they are carelessly handled. Punctuation, for example, plays a key role in written communication because it clarifies meaning and makes the reader's job easier.

UNIT 6.1 BASIC GRAMMAR

Grammar is a system of commonly accepted ways of forming words and word combinations to convey meaning. Most native speakers of a language automatically use the major conventions of its grammar because they have learned this system along with the language.

Speakers may also violate certain features of grammar without seriously disturbing the flow of meaning. They have a margin of safety because language is flexible and because speech communication includes many nonverbal signals such as gestures, tone of voice, and other clues. In writing, grammar becomes more important, not just for the sake of correctness but because in ungrammatical writing the meaning may not be clear.

Writing also gives you a chance to improve your first efforts. Check the first version to see that it says what you meant, and look at how it does so with a critical eye. In attempting to improve your first version, you may have questions about some of the points of grammar discussed in this unit.

The points included here are "basic" in the sense that they occur frequently in writing and stand out if they are not handled properly. For the writing you do in college, and on the job afterward, you should command at least this much grammar. These basic grammar principles include the following:

A. Sentence essentials
B. Sentence coherence
C. Subject-verb agreement
D. Pronoun forms, agreement, and reference
E. Plurals, possessives, and contractions

A. Sentence Essentials

As a writer, you must command the basic unit of meaning: the sentence. Writing sentences requires that you include all the essential elements and avoid two common pitfalls: (a) substituting fragments for sentences and (b) running several sentences together without separating them with punctuation.

☐ Write in sentences.

☐ Avoid fragments and run-ons.

A.1 Sentences and Fragments. A sentence is a complete thought. To complete a thought, you need at least a subject and a verb—an actor and an action. At the very least, the sentence may have only a verb, with the subject understood but not present. The following are all sentences:

Stop! (The subject *you* is understood.)

He stopped.

He stopped the engine.

Reaching into the desk drawer, he pulled out a thick file in a dull green binder.

A fragment, as its name suggests, is only a part of a sentence. A fragment may even contain a noun and a verb, but it is not a sentence because it does not complete the thought.

Sentence: She turned off the engine.

Fragment: After she turned off the engine.

Sentence: She reached into the desk drawer.

Fragment: Reaching into the desk drawer.

The most frequent cause of fragments in writing is the writer's failure to complete the thought within *one* sentence. There may be a completed thought, but it may be spread over two adjoining groups of words. Sometimes the thought is completed in the writer's mind but not on paper.

Wrong: She checked the fan belt. After she turned off the engine.

Right: She checked the fan belt after she turned off the engine.

Wrong: The feasibility report lay forgotten in a desk drawer. Bound in a dull green cover.

Right: Bound in a dull green cover, the feasibility report lay forgotten in a desk drawer.

Right: The feasibility report lay forgotten in a desk drawer. The report was bound in a dull green cover.

A.2 Sentences and Run-ons. Whereas the fragment is only part of a sentence, a run-on occurs when two or more sentences are run together as one. The most common type of run-on is the comma splice, in which two sentences are "spliced" together with only a comma instead of a period or semicolon.

Comma splice: A potential benefit of oligopoly is its ability to improve products and manufacturing techniques, the higher profits of oligopoly firms allow them to invest more in research and development.

Sentence: Oligopoly firms may be better able to improve their products; their higher profits leave them more resources for research and product development.

B. Coherent Sentences

Effective sentences are not only complete but also coherent. To write sentences whose parts work together instead of against each other, observe the following requirements:

☐ Avoid misplaced or dangling modifiers.

☐ Use parallel structure.

B.1 Misplaced or Dangling Modifiers. When a modifying phrase or clause is placed in the wrong part of a sentence, the results can be misleading:

Wrong: Wilson offered his proposal to the committee members *with pride.*

Wrong: *Conferring briefly,* the proposal was accepted by the committee.

The first sentence seems to indicate that only some members had pride; the second sentence has a proposal conferring briefly. These unintended meanings can be avoided by moving the two phrases next to the words they are intended to modify:

Right: *With pride,* Wilson offered his proposal to the committee members.

Right: *Conferring briefly,* the committee accepted the proposal.

Sometimes a modifier is not misplaced but is left dangling because there is no word in the sentence that it can logically modify (see Figure 6–1):

Wrong: *By replacing the spark plugs,* the engine ran smoothly again.

Because the engine didn't replace the spark plugs, the intended meaning must be expressed another way:

Right: *By replacing the spark plugs,* I got the engine running smoothly again.

To avoid misplaced modifiers, make sure the modifying phrase or clause is next to the word it is supposed to modify. To avoid dangling modifiers, make sure that the sentence actually contains the word to be modified and that the phrase or clause is next to that word.

B.2 Parallel Structure. Parallel structure is another way to make sentences more coherent and easier to read. When your sentence refers to a series of ideas or actions or things, put these elements into parallel form. The most common option you have is between the *-ing* form (*running*) and the infinitive form (*to run*):

Not Parallel: I like *to play* tennis and *swimming.*

Parallel: I like *to play* tennis and *to swim.*

Parallel: I like *tennis* and *swimming.*

FIGURE 6–1
When a modifying phrase is left dangling, unintended meanings can result.

A Dangling Modifier

By replacing the spark plugs, the engine ran smoothly again.

Parallel: I like *playing* tennis and *swimming* in the pool.

Elements that should be parallel may be joined by conjunctions such as *and, but, or, nor, either/or, not only/but also,* and so on. Lists should always have parallel form.

Not Parallel: This unit is not only well designed but compatible with our present equipment, as well as they offer a competitive price.

Parallel: This unit is not only well designed but also compatible with our present equipment and competitively priced.

Not Parallel: The group is going to the auditorium rather than try to meet in the seminar room.

Parallel: The group is going to the auditorium rather than trying to meet in the seminar room.

Parallel: The group is meeting in the auditorium rather than the seminar room.

Not Parallel: The meeting accomplished the following: revision of the project schedule, two prototypes were approved, and we modified the contract language.

Parallel: The meeting accomplished the following: revision of the project schedule, approval of two prototypes, and modification of the contract language.

Parallel: The meeting accomplished the following: We revised the project schedule, approved two prototypes, and modified the contract language.

C. Subject-Verb Agreement

Agreement between subject and verb is also needed to make the sentence coherent:

- ☐ Make verbs agree in number (singular or plural) with subjects.
- ☐ Make verbs agree in person (first, second, or third) with subjects.
- ☐ Remember that most subject-verb agreement problems occur only in the present tense.

C.1 Agreement in Number. Typically, agreement in number in the present tense means that singular subjects, which usually have no *s* ending, must be joined to singular verbs that usually have an *s* ending. For plural, it's just the reverse. Plural subjects, which usually have an *s* ending, are joined to plural verbs that usually don't have an *s* ending in the present tense.

Singular: A *circuit provides* a closed path through which the charge can flow.

Plural: *Circuits provide* a closed path for current flow.

Past Tense: The *circuit provided* a path.

The *circuits provided* a path.

C.2 Agreement in Person. Agreement in person is demonstrated by the combinations shown in Tables 6–1 and 6–2. In the present tense, most verbs require no more than the addition of an *s* ending to make them agree with subjects like *he, she,* and *it.* See Table 6–1.

The pattern differs for several short but important verbs, including *be, do, go,* and *have* (see Table 6–2). For these verbs, agreement in person follows the typical

TABLE 6–1
Agreement Between Person and Verb: Typical

Person	Singular	Plural
First	I take	we take
Second	you take	you take
Third	he/she/it takes	they take

TABLE 6–2
Agreement Between Person and Verb: Atypical

Subject	Singular				Subject	Plural			
	be	**do**	**go**	**have**		**be**	**do**	**go**	**have**
I	am	do	go	have	we	are	do	go	have
you	are	do	go	have	you	are	do	go	have
he/she/it	is	does	goes	has	they	are	do	go	have

pattern, except that the third person singular requires some atypical forms. In addition, *be*—the most contrary verb in English—follows its own pattern.

C.3 Other Subject-Verb Agreement Problems. C.3.a When the subject is separated from the verb by phrases, clauses, or interrupting words, make sure the verb agrees with the subject and not simply the nearest noun.

> The relative *advantages* of the front-loading model *are* easily demonstrated.
>
> The *source* of many of these customer complaints *is* our inefficient billing system.

C.3.b When the subject has several parts joined by *and,* make the verb plural; when a compound subject is joined by *or* or *nor,* make the verb singular.

> The *size and weight* of the new model *are* well within tolerances.
>
> Neither the *size nor* the *weight is* excessive.

C.3.c Subjects that are limited by words like *each, every, either,* and *neither* are singular. *Everyone, somebody, anybody,* and similar words are also singular.

> Neither *proposal has* much of a chance.
>
> *Everyone* who attends these seminars *comes* away impressed.

C.3.d A collective noun (*department, committee, team,* etc.) is generally singular and should be matched with singular verbs.

> The accounting *department is* headed by Monica Lazarde, a C.P.A.
>
> The executive *committee,* which met this morning, *has* approved the expansion.

Even though collective nouns refer to groups of people, usually they are thought of as singular because the group acts as a unit. Sometimes, though, the members of a group act as individuals, and the collective noun is therefore plural.

Plural: The *class are* receiving final grades while *they are* on vacation.

Singular: The *class meets* in Room 326 Tuesdays and Thursdays.

D. Pronouns

The following rules apply to the use of pronouns:

- ☐ Make pronouns agree with the words they refer to (their antecedents).
- ☐ Use a different pronoun form when the pronoun has a different function in a sentence.
- ☐ Make sure that pronouns refer clearly to particular antecedents.

D.1 Pronoun-Antecedent Agreement. Pronouns should agree in number, in gender, and in person with their antecedents.

Number: The board rejected both *proposals* because *they* were too expensive.

Gender: *Mrs. Fulton* told *her husband* that *his* business partner had called.

Person: Mr. Jonas, *I* am advising *you* to hire *her.*

Agreement in gender is a problem when the antecedent of a pronoun is singular but has no particular gender. Occupational titles (doctor, technician, systems analyst) or indefinite words like *somebody* can be either masculine or feminine.

> A doctor must use all [*his?/her?*] skill and experience to deal with the problems of this impoverished district.
>
> Somebody dialed my extension, but [*he?/she?*] hung up without speaking.

The best way to deal with the gender-agreement problem is to rewrite the sentence so you can use a plural pronoun or avoid the reference altogether:

> Doctors must use all *their* skill and experience. . . .
>
> Somebody dialed my extension but hung up without speaking.

The combination *his or her* or *he or she* can also be used but is more awkward and should be minimized. The worst solution is to use a plural pronoun to refer to a singular antecedent—for example, "Somebody left *their* calculator on the counter." This is clearly illogical because "somebody" is one person. Use "Somebody left *a* calculator" instead.

D.2 Pronoun Forms and Functions. Pronouns serve three basic functions in a sentence: They can be subjects, objects (of a verb or preposition), or possessive modifiers. For each of these functions, use the forms of the pronouns shown in Table 6–3.

> *She* [subject] asked *him* [object of *asked*] about *his* [possessive modifier] views of *their* [possessive modifier] meeting with Finchley.
>
> Ms. Jones and *I* [subject] narrowed the choice to Mr. King and *her* [object of *to*].
>
> Just between you and *me* [object of *between*], Carla and *he* [subject] are the only serious candidates.
>
> The auditors objected to *our* [possessive modifier] claiming the trip as a business expense. [*Note:* The auditors objected to our action of claiming, not to *us.*]

D.3 Relative Pronouns. The relative pronouns *who, whom, whose, which,* and *that* occur in clauses within a sentence or begin questions. The choice of these forms is dictated by their function (subject, object, possessive) or by their reference to a person rather than a thing or idea. *Which* and *that* are used to refer to things or ideas: "The car *that* passed us . . . " "The report, *which* filled 33 pages, was delivered Friday." The possessive form *whose* can refer to any logical antecedent: "The student *whose* book . . . " "The package *whose* weight . . . "

The choice of the relative pronoun is thus a problem only between *who* (the subject form) and *whom* (the object form):

> *Who* took my newspaper? (subject)
>
> *Whom* can you trust? (object)
>
> The district manager, *who* flew in last night, is leaving this afternoon. (subject of clause)
>
> The district manager, *whom* we invited for the conference, is flying in tonight. (object of clause)
>
> Three of the applicants to *whom* we made offers accepted. (object of *to*)
>
> The job will go to *who*ever scores highest on the exam. (subject of clause, not object of *to*)

TABLE 6–3
Pronoun Forms

Subject		Object		Possessive	
I	we	me	us	my	our
you	you	you	you	your	your
he/she/it	they	him/her/it	them	his/her/its	their

D.4 Clarifying Pronoun Reference. Unclear pronoun reference occurs when a pronoun can be taken to refer to several antecedents or to a general idea:

Unclear: Rahill told Bradley that *he* would be in charge of the project.

(Is *he* Rahill or Bradley?)

Unclear: Three of the components arrived by parcel post Monday. *This* upset Linda.

(Does *this* refer to only three components? To arrival by parcel post? Arrival Monday? All of the above?)

In cases like the preceding, the unclear reference can be eliminated by (1) naming the antecedent, (2) using a general category noun next to the pronoun, or (3) revision to eliminate the pronoun:

Clear:

1. Rahill told Bradley that Bradley would be in charge.
2. Three of the components arrived by parcel post Monday. The late arrival upset Linda.
3. a. Rahill put Bradley in charge of the project.
   b. Linda was upset because three of the components arrived by parcel post as late as Monday.

E. Plurals, Possessives, and Contractions

Use the following guidelines to form plurals, possessives, and contractions:

- ☐ Form most plural nouns by adding *s* or *es.*
- ☐ Never use *'s* to form plurals.
- ☐ Use *'s* to make words possessive. If the word is a plural that ends in *s,* just add the apostrophe ('). For proper names, including those ending in *s,* add *'s.*
- ☐ Use the apostrophe to form contractions.

E.1 Noun Plurals. Most nouns can be made plural by simply adding an *s* ending (*project, projects; system, systems*). Other ways of forming noun plurals include the following:

- ☐ tax/taxes, cross/crosses (Add *es* for nouns ending in *s, z, x, ch,* or *sh.*)
- ☐ fallacy/fallacies, attorney/attorneys (For nouns ending in *y,* change *y* to *i* and add *es,* unless the *y* is preceded by a vowel.)
- ☐ shelf/shelves, half/halves, roof/roofs (For *f* endings, change *f* to *v* and add *es* in most cases.)
- ☐ radio/radios, video/videos, cargo/cargoes (For *o* endings, add *s* in most cases.)
- ☐ tooth/teeth, man/men (Internal vowel change.)

- ☐ analysis/analyses, hypothesis/hypotheses (For words of Greek origin, change *is* to *es.*)
- ☐ nucleus/nuclei, stimulus/stimuli (For words of Latin origin, change *us* to *i.*)
- ☐ datum/data, medium/media (For words of Latin origin, change *um* to *a.*)
- ☐ criterion/criteria, phenomenon/phenomena (For words of Greek and Latin origin, change *on* to *a.*)
- ☐ series/series, species/species (Same in singular and plural.)
- ☐ music, electronics, mathematics (No plural form.)
- ☐ scissors, slacks, barracks (No singular form.)
- ☐ editor-in-chief/editors-in-chief, quarter-final/quarter-finals (Change the main word of a compound noun.)

Check a dictionary for any noun plural forms you are not sure of.

Never use *'s* to form a plural, not even for numbers, letters, symbols, or short words referred to as such. Just add *s.*

the 1990s
The VCRs were delivered.
three 9s in the answer
too many *of*s in this sentence

E.2 Possessives. Show possession by adding *'s* to nouns; add only an apostrophe to nouns already ending in *s.* For proper names, add *'s.*

modem/modem's
modems/modems'
series/series'
people/people's
Lucy/Lucy's
Charles/Charles's

Because pronouns already have possessive forms (*his, its, ours*), there is no need to add *'s* or an apostrophe. The only pronouns that take the *'s* ending are the indefinite forms, such as *someone* or *anybody* (see Table 6–4).

E.3 Contractions. Contractions are formed by omitting part of a two-word combination (we will) and substituting an apostrophe for the omission (we'll). Some pronoun-verb and verb-modifier combinations are contracted frequently in informal writing or speech (it is/it's; do not/don't). Contractions are used less frequently in business and technical writing.

TABLE 6–4
Possessive Pronouns

Personal		Relative	Indefinite
Singular	**Plural**		
my/mine	our/ours	whose	everyone's
your/yours	your/yours		anyone's
her/hers			someone's
his	their/theirs		no one's
its			everybody's
			anybody's
			somebody's
			nobody's

Whether used frequently or sparingly, contractions should not be confused with possessive or plural forms, which sometimes have similar features.

Singular	*Plural*	*Possessive*	*Contraction*
system	systems	system's/systems'	system's (system *is*)

The following distinctions are frequently troublesome:

Possessive	*Contraction*
its	it's (it is)
whose	who's (who is)
your	you're (you are)
their	they're (they are)
hers	here's (here is)

The most common possessive-contraction error is to confuse *its* with *it's.* You can avoid the confusion by "uncontracting" the term (*it's* = *it is*) and then checking whether the usage still fits.

Right: The plane lost *its* left engine during takeoff.

Wrong: The plane lost *it's* [it is] left engine.

Unit 6.1 Exercise

A. Identify the following items as *sentences, fragments,* or *run-ons* by entering *S, F,* or *R* in the blank spaces.

1. _____ Working on the project through the Labor Day weekend while everyone else was at the beach.

2. _____ I couldn't get the figures to agree, the error must be in the original.

3. _____ Be careful.
4. _____ Take the I-435 bypass, it's faster.
5. _____ A revised edition, which has case studies and discussion questions after each chapter.
6. _____ After the storm dumped eight inches of fresh snow on the north end of the city, schools closed.
7. _____ At two of the locations, access roads for delivery trucks had to be built.
8. _____ At the third location, which, in fact, received the highest rating for favorable demographic factors in the study we conducted.
9. _____ They decided to make the course a graduation requirement, it's already a prerequisite for the accounting sequence.
10. _____ His tendency to drop by the office at closing time for extended reviews of the problems of the day.

B. Rewrite the following sentences to eliminate the misplaced modifiers.

Example:

Wrong: In a show of hands, the remaining proposals were adopted by the committee.

Right: In a show of hands, the committee adopted the remaining proposals.

1. Ordering six weeks before the due date, the parts were received in plenty of time by Thelma.

2. A new sedan was unveiled at the fall auto show with front-wheel drive.

3. Randy diagnosed the problem as a twisted cable, hoping he was right.

4. Sweeping the field with binoculars, a slight movement in the hedgerow drew Millicent's attention.

5. The inspector found hairline cracks in the foundation that hadn't been there yesterday.

6. Refusing to concede any of their advantages, the problem remained deadlocked between the two sides.

7. By carefully positioning the ceiling mirrors, the back of the store was visible from the front counter.

C. Rewrite the following sentences to establish parallel structure.

Example:

Not Parallel: The speech was logical, well developed, and persuasion was used effectively.

Parallel: The speech was logical, well developed, and persuasive.

1. Check the oil, rotate the tires, and the air filter should be replaced.
2. We must either hire a new secretary or Gary can be asked to work overtime for two weeks.
3. This package is not only more powerful but also using it is easier.
4. I stayed with my cousin, an electrical engineer and who is a gourmet cook.
5. I liked the movie for its characters, its plot, and the cinematography was outstanding.
6. The more I practiced this technique, I enjoyed skiing more.
7. It may be more productive to go back to the beginning than trying to find the way out from here.

D. Make the verb agree with the subject by circling one of the terms in brackets.

1. Repairs on the snow plow [was/were] expected to take two days.
2. Surveying the final sector and writing the summary report on the project [has/have] kept the team busy.
3. Because of the supply problems, either Tom or Trudy [is/are] going to be assigned to the Phoenix facility next month.
4. The most frequent complaint we found on these comment cards from first-time users [concerns/concern] the installation instructions.
5. Each of the shipments to the overseas customers [requires/require] special handling.
6. Everybody who took the refresher courses [has/have] found them worthwhile.
7. In that crowded primary field of familiar names and famous faces, neither candidate from the two major parties [was/were] able to establish a clear lead.

8. The mood of the voters in the seven primary states that go to the polls in the next several weeks [remains/remain] largely mysterious.
9. A research paper and a major design project [is/are] required in the course.
10. During the last stage of the trip, the rolling countryside and the swaying motion of the car always [makes/make] me sleepy.

E. In each sentence, underline the antecedents, and circle the pronouns that agree with the antecedents.

1. The last of the passengers took [his/their] seat, and the train began to move.
2. Deborah felt pleased about [she/her] standing in the class after [it/they] selected her Student Council representative.
3. Bill said that George and [he/him] were familiar with the problem and would deal with [it/them] as soon as possible.
4. Grigsby told Haldane, "Just between you and [I/me], neither of the auditors thought [she/they] could accept the deduction."
5. Petrakis realized early on that the success of the plan depended to a great extent on [he/him/his] staying clear of the implementation details.
6. Harriet looked forward to the appointment of Wilbur and [she/her] to the commission.
7. Mr. Philby, [who/whom] I called this morning, said he was satisfied with the new equipment.
8. The amended service contracts to [what/whom/which] she referred had been lost in our move to the Oakton office.
9. She is one of the best pianists [who/whom/that] we have ever heard at Orchestra Hall.
10. The scholarship will go to [whoever/whomever] writes the best essay on the potential of electronic communication.

F. In each sentence, underline the pronoun whose reference is unclear. Then rewrite the sentence to eliminate the unclear reference.

1. The student told the professor that she had been mistaken.

2. There was a long discussion on computer modeling; this bored Jessica.

3. In the *Times,* it says good tickets for the concert are still available.

4. The systems department office is located in the annex building; it is being remodeled today.

5. The surge in consumption overloaded the system, which led to the power outage.

6. Since the city council passed the snow ordinance, they have been towing cars on our street every time a few snowflakes fall.

7. After unbolting the metal plate with a wrench from his tool kit, he put it on the counter.

G. Write the plural forms of the following nouns:

1. bench ____________	6. analysis ____________
2. stereo ____________	7. display ____________
3. passerby ____________	8. leaf ____________
4. foot ____________	9. criterion ____________
5. company ____________	10. series ____________

H. In each sentence, underline any words that incorrectly use possessives, contractions, or plurals. Rewrite the underlined forms correctly.

1. The attorney's filed their briefs in the court clerk's office.

2. This job is our's if we can beat one other companies' bid.

3. I'm not sure who's responsibility this contract is, but everyones concerned.

4. This car's probably on it's last trip before it'll need a major tuneup.

5. The marching bands forming in the southwest corner of the parade ground's.

6. The Hartley's summer vacation trip to their best friend's, the Jolsons, was postponed.

7. The shipments been delayed at they're end by a trucker's strike.

8. Copies of all the committee members assignments are in you're desk drawer.

TABLE 6–5
Verbs With and Without Objects

Present Tense	Past Tense	Present Participle	Past Participle
	Verbs without objects		
rise	rose	rising	risen
sit	sat	sitting	sat
lie	lay	lying	lain
	Verbs with objects		
raise	raised	raising	raised
set	set	setting	set
lay	laid	laying	laid

9. Wilsons going to keep his territory and add her's.

10. After it's final field test, all the medias should be contacted about the introduction at the March trade show.

UNIT 6.2 VOCABULARY AND USAGE

Two categories of vocabulary and usage problems are discussed in this unit: (a) words that are frequently confused with each other, and (b) words whose meaning does not fit the intentions of the user.

A. Frequently Misused Word Pairs

The pairs or groups of words in this section are frequently confused with one another. The source of the confusion may be a similarity in spelling or pronunciation, a lack of precision in meaning, or a combination of such factors.

Use pronunciation and spelling differences as clues to which term should be used. Check the following explanations and examples to determine the meaning of a term. For other terms you are not sure of, consult a college-level dictionary.

(See Table 6–5 for the forms of *rise/raise, sit/set,* and *lie/lay.* These are among the most troublesome pairs to distinguish.)

accept/except The verb *accept* means to receive or to agree to. The preposition *except* means "with the exclusion of."

> We *accept* the terms of the contract as specified, *except* for the two clauses in fine print.

advice/advise The noun *advice* refers to recommendations or suggestions. The verb *advise* means to recommend or to suggest.

> The consultants *advised* further study, but Lawrence was impatient and rejected their *advice.*

affect/effect The verb *affect* means to influence. The noun *effect* means "a result."

> The budget cutbacks could *affect* several critical areas; the *effect* on research and development, for example, may be quite severe.

all ready/already *All ready* means "completely prepared." *Already* means "before this time."

> The product had *already* been on the market for five months before our service centers were *all ready* to provide maintenance.

among/between Use *between* to relate two things; use *among* for more than two.

> Her report analyzed the key differences *between* the two models; these models were the best *among* the five we were considering.

amount/number Use *amount* with things that are not counted individually; use *number* for things that are counted.

> The *amount* of time spent on researching companies should be proportional to the *number* of job offers received.

(Note: The number of hours may be counted, but time is uncountable.)

are/or/our Check pronunciation and spelling to help distinguish the verb *are,* the conjunction *or,* and the possessive pronoun *our.*

> *Our* new neighbors *are* dropping by Wednesday *or* Thursday.

beside/besides *Beside* means "next to"; *besides* means "in addition" or "except."

> The van stood *beside* the garage, and *besides,* there was nothing in it *besides* a broken chair.

brake/break As nouns: a *brake* is a mechanism for slowing or stopping, and a *break* is a rupture. As verbs: to *brake* is to slow or stop, and to *break* is to rupture or destroy.

> A *break* in the fluid line to the *brake* led to the crash.

bring/take *Bring* means to carry toward; *take* means to carry away from. Don't use *bring* for *take.*

> *Take* this set to the service department, and *bring* back a copy of the repair order.

can/may *Can* means "able to," and *may* means "allowed to" or "free to"; *may* is also used to indicate probability. Don't use *can* for *may.*

> The publisher said we *may* use the article if we *can* assure him it's for educational purposes; I *may* send him a letter today.

capital/capitol The *capital* is the city; the *capitol* is the legislative building.

> Flights to the state *capital* were crowded because of tomorrow's ceremonies on the *capitol* steps.

clothes/cloths *Clothes* are garments you wear; *cloths* are pieces of fabric used for cleaning and other odd jobs.

> Put on your work *clothes* while I find some clean *cloths* for washing the car.

coarse/course *Coarse* means "rough" or "crude"; a *course* is a class in school.

> I brought some *coarse* sandpaper for my woodworking *course.*

complement/compliment A *complement* balances or completes something; a *compliment* is praise.

> She earned a *compliment* from the math teacher for pointing out that angle ABC is the *complement* of angle CBD.

continual/continuous A *continual* process has interruptions; a *continuous* process doesn't.

> During the meeting, there were *continual* complaints about the *continuous* whine of the air conditioner.

criteria/criterion *Criteria* is the plural of *criterion.* Don't use the form *criterias.*

> We have several *criteria* for selecting a contractor, but our main *criterion* is quality of workmanship.

data/datum *Data* is the plural of *datum.* Don't use the form *datas.*

> Except for a single *datum* involving piston wear, the *data* on reliability have been assembled.

device/devise The noun *device* means a mechanism or a scheme; the verb *devise* means to invent or to design.

> We have to *devise* a method for bypassing the *device* that shuts down the assembly line.

discreet/discrete *Discreet* means tactful; *discrete* means separate and distinct.

> He made several *discrete* changes, but his comments about the report were very *discreet.*

each other/one another Use *each other* for two-party relationships; use *one another* for groups of three or more.

> The engineer and the claims adjuster supported *each other,* but the rest of the jury members were at odds with *one another.*

ensure/insure *Ensure* means to make certain; *insure* means to guard against financial loss.

> To *ensure* their acceptance of the contract, we will have to *insure* our delivery vehicles.

farther/further *Farther* means "a longer distance"; *further* means additional.

> We had no *further* contact with Bill and Beverley after they moved *farther* from town.

fewer/less Use *fewer* with countable nouns; use *less* with uncountable nouns.

> With *fewer* committee members present, there was *less* dissension during the meeting.

FIGURE 6–2
Good speakers distinguish *good* and *well*.

forth/fourth *Forth* means "out from" or "forward"; *fourth* means "number four."

On the *fourth* day, a witness came *forth* to testify.

good/well Use *good* with nouns, *well* with verbs. But use *good* after verb forms of *to be* (*is, are, was,* etc.) and after verbs such as *look, taste, feel, smell, seem.*

They're a *good* team; they work *well* together. The service here is *good;* my car looks *good,* sounds *good,* and handles *well.*

have/of Don't use *of* for the verb *have* in combinations such as *would have* or *should have.*

hear/here To remember *hear,* think of your ear. *Here* is a place.

Did you *hear* that the conference will be *here* in town?

idea/ideal An *idea* is a thought; *ideal* refers to perfection.

It's a good *idea,* but it's not the *ideal* solution.

imply/infer To *imply* is to suggest; to *infer* is to deduce or figure out. Don't use *infer* for *imply.*

Their ads *imply* that they're a good corporate citizen; we can *infer* that the company is concerned with its image.

in/into *In* means inside or within; *into* refers to movement from outside to inside.

I was *in* the lobby when they came *into* the building.

inter-/intra- Use the prefix *inter-* to suggest *between;* use *intra-* to suggest *within.*

There was an *inter*national conference on the *inter*state regulation of *intra*venous medicines.

its/it's The possessive *its* ("belonging to it") doesn't need an apostrophe; *it's* is a contraction for *it is.*

It's too late now to alter *its* orbit.

knew/new *Knew* means "had knowledge"; *new* means recent or modern.

We *knew* that the *new* orientation program would take time to establish.

lay/lie *Lay* is used with objects; *lie* has no object. Note also that the past form of *lie* is *lay* (see Table 6–5).

I asked the repairman to *lay* the manual on the shelf, but he let it *lie* on the floor. It *lay* there in the dust all afternoon.

leave/let *Leave* means to depart or to allow to remain; *let* means to permit. Don't use *leave* for *let.*

> *Let* me suggest that you *leave* the plans on your desk when you *leave.* I will study them and *let* you know tomorrow.

liable/likely *Liable* means obligated or responsible in a legal sense; *likely* means probable.

> We are *likely* to be held *liable* for damages in the antitrust case.

loose/lose *Loose* means unattached; *lose* means to misplace or to be deprived of.

> Don't *lose* your grip; if the cable comes *loose,* we may *lose* the cargo.

may be/maybe *May be* means "might be"; *maybe* means "possibly."

> There *may be* a way to fix this printer, or *maybe* we should buy a new one.

media/medium Because *media* is the plural of *medium,* don't use the form *medias.*

> Among the *media* of communication, her favorite *medium* is still the newspaper.

moral/morale A *moral* sense distinguishes right and wrong; *morale* is the way people in a group or organization feel.

> Department *morale* improved after the *moral* stance taken by the company.

passed/past The verb *passed* means "went by"; the noun *past* refers to the time before the present.

> This company was content to rest on *past* accomplishments until the rest of the industry *passed* it by.

percent/percentage Use *percent* with a number; use *percentage* when a number is not specified. (In technical writing, you will commonly use the symbol % with numbers.)

> Less than 5 *percent* of the questionnaires were completely filled out, but a much higher *percentage* included comments at the end.

personal/personnel *Personal* means private; *personnel* refers to employees.

> The *personnel* policies here restrict use of telephones for *personal* calls.

precede/proceed To *precede* is to go before; to *proceed* is to continue.

> The bus driver wanted to let the taxi *precede* him, but the policeman waved at the bus to *proceed.*

principal/principle *Principal* means "main" or "primary"; a *principle* is a standard or general truth.

> His *principal* contribution to the design was a feedback loop based on the Murphy *Principle.*

quiet/quite *Quiet* means silent or peaceful; *quite* means "to a great extent."

> Setting up a *quiet* study area was *quite* important to Dean Swift.

raise/rise *Raise* has an object; *rise* doesn't (see Table 6–5).

> If our costs *rise* again, we'll have to *raise* our prices.

real/very *Real* means genuine; *very* means extremely. Don't use *real* for *very.*

> My partner did a *very* good job on this project; his research was a *real* contribution to our success.

right/write *Right* means correct; to *write* is to communicate.

> I wanted to *write* her a note to tell her she had been *right* about the situation.

set/sit *Set* has an object; *sit* doesn't (see Table 6–5).

> Let's *set* a time limit for this meeting before we *sit* down at the table.

sometime/sometimes *Sometime* refers to an unspecified time in the future; *sometimes* means "now and then."

> *Sometimes* these units arrive damaged; we should have a long talk with shipping *sometime.*

than/then *Than* means "compared with"; *then* refers to a particular time or time period.

Then he pointed out that the machine was bigger *than* the shipping crate.

their/there/they're *Their* is the possessive pronoun; *there* refers to place; *they're* is a contraction for *they are.*

Their expansion plan suggests *they're* doing well *there.*

to/too/two *To* indicates approach; *too* means also or more than desired; *two* is the number.

Then *too,* it's *too* far *to* Toronto for a *two*-day trip.

weather/whether *Weather* refers to climate; *whether* means "if".

As the *weather* worsened, it became less clear *whether* the launch would go off on schedule.

who's/whose *Who's* is the contraction for *who is; whose* is the possessive pronoun.

Whose car we use depends on *who's* going on the trip.

your/you're *Your* is the possessive pronoun; *you're* is the contraction for *you are.*

You're going to be called as soon as *your* number comes up.

B. Frequently Misused Words. Following is a group of terms that pose usage problems. The problems occur in both writing and speech and persist from habit. For most of these terms, a logical distinction can be made between good usage and poor usage.

a lot Should always be written as two words, and never as *alot.*

aggravate Means to make worse. It should not be used for *irritate* or *annoy.*

I know that if I get *annoyed* and complain, I will *aggravate* the problem.

as to, as far as . . . is concerned Avoid these vague expressions. Use *about* or *with*

Vague: They're curious *as to* the early results of the experiment.

Vague: They're curious *as far as* the early results *are concerned.*

More Precise: They're curious *about* the early results.

FIGURE 6–3
While *you're* waiting for *your* number to come up, you might as well sit down.

due to Should be used only after the verb *to be;* otherwise, use *because of* or *resulting from,* as appropriate.

> *Because of* the inventory shortfall, we rescheduled milling operations; the shortfall was *due to* a strike in Pennsylvania.

et al., i.e., e.g. The Latin abbreviation *et al.* means "and others" and is commonly used in referring to several authors (e.g., Wozniak et al.). Notice there is no period after *et,* which means "and." The abbreviation *i.e.* means "that is," and *e.g.* means "for example"; they should be used only in parentheses. When used in text they should be spelled out in English: *id est* and *exempli gratia.*

> They want delivery by the 20th (*i.e.,* the impossible).
>
> Several companies (*e.g.,* TVM and Decker) are interested.

expect, suppose *Expect* means anticipate or await; *suppose* means to presume or think likely. Don't use *expect* for *suppose.*

> I *suppose* they will *expect* a discount on this model.

former, latter The terms *former* and *latter* refer to the first and last of *two* named things. With more than two, use *first* and *last.*

> Yesterday I finished the introduction and the body of the report; the *former* took me longer than the *latter.*

hardly, scarcely Don't use *not* with *hardly* or *scarcely.* These are negative terms already and become double negatives when *not* is added.

> *Wrong:* We could*n't hardly* see the road in the fog.
> *Right:* We could *hardly* see the road in the fog.

hopefully Means "with hope" and shouldn't be used to say "it is hoped" or "let's hope."

> *Wrong:* *Hopefully,* the weather will improve tomorrow.
> *Right:* Let's hope the weather improves tomorrow.

irregardless Why add *ir-* when the word already ends in *-less?* Just use *regardless.*

> He wanted a McIntosh coat, *regardless* of cost.

the reason is . . . Avoid redundant phrases such as "the reason is because" or "the reason why"; use *the reason is* with *that,* or avoid the phrase altogether.

> *The reason is that* the power failed.
> *The reason* she was late *is that* her car failed to start.
> She was late *because* her car failed to start.

that, which *That* starts a clause essential to the meaning of previous words and, therefore, not set off by commas; *which* is used in a nonessential clause.

> The suggestion *that* we finally adopted came from Andrea.
> Andrea's suggestion, *which* we finally adopted, came near the end of the meeting.

try to Do not replace *try to* with *try and.* "I will *try to* solve the problem" sets up the verb *solve.* "I will *try and* solve . . . " sets up two verbs (*try, solve*) and suggests that I "will solve" instead of only "try."

use, utilize These words mean the same thing, so *use* the shorter one.

Unit 6.2 Exercise

A. In the following sentences, circle the correct word in brackets.

1. I was astonished at the [amount/number] of [intermural/intramural] teams fielded by our department.
2. Gunther is doing [good/well] at school, and he feels [good/well] about his career prospects.
3. The [raise/rise] in producer prices could [affect/effect] the economy in the last quarter.
4. [Can/May] we study in this empty classroom to get away from the [contin-ual/continuous] interruptions in the commons area?
5. A very high [percent/percentage] of the new models experienced stalling problems in wet [weather/whether].
6. He let the report [lay/lie] on his desk all week; it has [laid/lain] there long enough.
7. The [forth/fourth] problem took longer [than/then] I expected.
8. The [principal/principle] objection to the plan was [its/it's] cost.
9. The vote was evenly split [among/between] the three candidates who spoke from the [capital/capitol] steps.
10. The new advertising campaign attracted [fewer/less] customers, but sales figures are [liable/likely] to be higher.
11. We should locate the copier [farther/further] from the door to [ensure/insure] a constant temperature for operations.
12. You can't just [set/sit] there and hope [your/you're] going to [hear/here] from clients.
13. The car lost [its/it's] [breaks/brakes] during the trip to Phoenix.
14. If we can [accept/expect] these terms, repairs on this [device/devise] will begin next week.
15. Writing a paper seems like a [real/very] attractive alternative, one that [complements/compliments] the strong quantitative focus of the course.
16. These endless delays are having a negative [affect/effect] on [moral/morale].
17. Their attorney's comments [implied/inferred] that they would be willing to [accept/except] our company from their complaint about [personnel/per-sonal] practices in the industry.
18. He had already [raised/risen] from his chair, so the knock on the door [passed/past] unanswered.
19. I'm not sure, but this [may be/maybe] the part [their/there/they're] look-ing for.
20. And now, let's [precede/proceed] with a discussion of selection [criteria/criterias] for the [ideal/idea] candidate.

B. Correct any errors in usage in the following sentences:

1. There will, hopefully, be a good turnout for the meeting tomorrow.
2. He was so aggravated by the questioning, he stalked off the platform.
3. Due to the early frosts, produce will be more expensive this winter.
4. They are very disappointed as far as this antitrust case is concerned.
5. I expect you will want to see the written report before you decide.
6. Give your work experience in reverse chronological order (e.g., in reverse order of occurrence).
7. Alot of the students left early for the holidays.
8. Joan's oral presentation, that she had practiced delivering for several days, was very well received.
9. The room was so large, I couldn't hardly see the diagrams he was drawing on the board.
10. We should try and make a damage assessment before the press conference.
11. Bring me the manual which gives the assembly instructions for this model.
12. The position offers several unusual benefits i.e., a polo club membership and a free subscription to *Sport of Kings* magazine.
13. The reason the power went out is because a fire broke out in the switching station in Decatur.
14. Why don't we utilize the boardroom for the meeting on Friday?
15. *The Case for Artificial Intelligence,* by Przybski et. al, was placed on reserve in the library.

UNIT 6.3 CAPITALIZATION

Words are capitalized to indicate that a specific name is being used. Thus, the specific names of specific persons, places, things, or ideas are capitalized, and the general categories of persons, places, things, or ideas are written in the lowercase.

General	***Specific***
inventor	Thomas A. Edison
skyscraper	Sears Tower
personal computer	Macintosh
department	Graduate Placement Department
company	Ford Motor Company
book	*Brave New World*
mountain	Pike's Peak
idea	Ohm's Law

Don't capitalize general category names unless they are part of a specific name:

A key *battle* in World War II was the *Battle* of the Coral Sea.

On the trip we saw *lakes* Michigan and Superior, but we bypassed *Lake* Erie.

We took them to *court,* and the case ended up in the *Supreme Court.*

Looking for an empty *room,* he walked into *Room* 31 during a class.

I bought a loaf of Wheatstone *bread* and a Royal *personal computer.*

Capitalize adjectives derived from the names of countries, races, nationalities, religions, political groups, and other such communities:

Italian cooking

English culture

African art

Caucasian features

Islamic principles

Republican candidate

Don't capitalize job or occupational titles unless they directly precede the person's name:

He introduced Ruth Radcliffe, *president* of the Haverford Institute, to *President* Harbrace of Exeter College.

She recommended I see a *doctor,* so I called *Dr.* Turner.

At *Dean* Kurlandski's request, we surveyed all the *deans* and *professors* in the college.

Capitalize specific academic degrees, whether abbreviated or written in full; don't capitalize general categories of degrees such as *bachelor's* or *master's:*

Dr. Ralls got his *bachelor's* in chemistry and then earned an *M.A.* and a *Ph.D.* in educational administration.

Capitalize the names of weekdays, months, and holidays. Don't capitalize seasons:

His birthday is next *Monday,* which is *Washington's Birthday.*

The *fall* semester began right after *Labor Day.*

Capitalize the names of defined geographical regions but not compass directions:

We drove *south* from Vancouver into the *Pacific Northwest.* In a diner just *east* of Portland, we met a traveler from the *Panhandle* of Texas.

Capitalize the specific names of school courses but not general academic fields, unless these also refer to languages or nationalities. Capitalize the names of departments or divisions of companies but not the names of general vocational areas:

I had math, physics, English, history, and psychology in high school. This term I'm taking Calculus I, Systems Analysis II, Managerial Accounting, and Contemporary History.

The financial analysts in Accounts Receivable are attending a seminar on edp auditing.

Capitalize the first word of a sentence and the first word of a formal statement after a colon. Capitalize the first word and all major words in titles of written works and the names of periodicals, plays, songs, movies, and shows:

The company released the following statement: We believe the article entitled "Dim Prospects at Highgate" in Monday's issue of the *Wall Street Journal* contains significant factual errors.

On the train I read Finchley's report on "Cash Flow in the Small Appliance Division," several chapters in *The Third Wave,* and an advertisement for the movie *A Dangerous Season.*

Capitalize people's titles and the names of family relationships when these directly replace a person's name.

Good morning, *Doctor.* This is my *cousin* Eileen.

My *father* was late, so *Mother* negotiated with the contractor.

Unit 6.3 Exercise

In the following sentences, circle any letters that should be capitalized, and draw a slash through any letters that should be lowercase.

1. after he read the Article in *consumer reports,* sam visited three Dealers to price a chrysler minivan.
2. she prepared her Report on a compaq prolinia personal Computer.
3. the Subway entrance was at the corner of state and madison Streets.
4. in the first debate, the democratic Candidate was chosen by lot to make the opening statement to the audience in burton auditorium.
5. weakening Markets in europe and south America were beginning to affect the japanese Economy.
6. my Appointment is in the wrigley building, which is that graceful white building across the River.
7. now that she was a bank Vice President, she began to sympathize with vice president Nelson of the first national bank, who was always being pursued for investment advice by his neighbors.
8. at the Reception, I ran into dr. Washburn, my Dentist.
9. the invitation list had failed to include the President of the Cambridge supply company, our new subcontractor.

10. Norman decided to pursue a Master's in electrical Engineering instead.
11. later she enrolled in the M.B.A. Program at the same University where she had earned her Bachelor's.
12. this Project is due at the end of the Spring semester.
13. this year, independence day falls on a Monday, so we'll have a three-day Weekend for the trip to grandma's.
14. to get to the Park entrance, you have to go Northwest from here for about five miles.
15. this will be her first trip to the west coast since June of last year.
16. my adviser said that if I want to major in Business, I'll need to take more courses in Economics, Accounting, Science, Algebra, and Spanish.
17. the new class schedule I worked out includes cost accounting II, college algebra, and data base management systems II.
18. we're going to meet with two of the Analysts in the marketing research department to discuss the Survey results.
19. the press secretary issued the following statement: the Administration believes that the Article in the *new york times* was premature and based on erroneous Information.
20. professor carter assigned Chapters 2 and 3 in addition to the Stoner article, "a somatic mutation theory," in the *journal of psychosomatic research.*
21. I told the Lieutenant that captain abrahams wanted to see him.
22. she brought along her Uncle, walter mitty, to meet aunt Alice.
23. she kept saying, "Oh, mother," all through dinner.
24. they planned a Canoe Trip on the chain of lakes at the Canadian Border.
25. next saturday is the Anniversary of the mount saint helens Eruption.

UNIT 6.4 PUNCTUATION

Punctuation clarifies meaning by indicating the boundaries of sentences and the relationships between words within the sentence. Punctuation marks are the road signs that help a reader navigate a passage of writing according to the writer's instructions.

Period (.)

Use a period at the end of a sentence and for abbreviations. If a sentence ends with an abbreviation, use only a single period.

> Dr. George L. Aubert, Jr., set up a display of vintage cars in the lobby of Highgain, Inc.

Question Mark (?)

Use a question mark after a direct question but not after a statement *about* a question.

> Who wrote this manual?
>
> I wonder who wrote this manual.

Exclamation Point (!)

Use an exclamation point to indicate strong emotion such as surprise, urgency, shock, or determination. Do not overuse it, or it will lose its effect.

> Don't touch that switch!
>
> To build this prototype, we would need more employees than the government and more money than the government owes!

Comma (,)

Use commas to separate words and to establish relationships between groups of words. Generally, a comma indicates a change in the direction of the thought. A comma never indicates a stop. (To put a "stop" to a train of thought, use a period or a semicolon.)

Comma Rule 1. Use a comma after an introductory dependent clause, after a long introductory phrase, and sometimes after an introductory word.

1a. Clause:

> Although it costs us more to serve some users, we are required to charge the same rate to all.

1b. Long phrase:

> In addition to the major types of analog modems, we discussed several other analog signal-conversion devices.

1c. Word:

> Recently, digital signal processors have been a topic of conversation in the department.
>
> *Note:* In 1c, the comma is used to avoid the false start of "recently digital."

Comma Rule 2. Use a comma *and* a conjunction (*and, but, or, so, yet*) to join two independent clauses.

> Most robot systems today are used in manufacturing, but their potential may be greatest in the service industries.

Comma Rule 3. Use a comma to set off a dependent clause following an independent clause if the second clause provides only supplementary information for the meaning of the first clause. When the second clause is essential to the meaning of the first clause, don't set it off with a comma.

We charged the same rates to all customers, although some lived outside our service area.

We charged the same rates to all our customers because we wanted to comply with ICC requirements.

We charged higher rates to customers who lived outside our service area.

Note: No comma is used after *customers* in the third example because *who lived outside our service area* is needed to identify which customers get charged more.

Comma Rule 4. Use a comma to separate parenthetical expressions (interrupting words) from the rest of the sentence. The parenthetical expression may be a word, a phrase, or a clause that provides additional information. If the parenthetical information is essential to the meaning of the sentence, don't set it off with commas.

3a. Word:

Data processing management, however, must tell us that the system is feasible.

3b. Phrase:

Corporate management, on the other hand, must weigh the costs of the new system against the promised savings and increased levels of information.

Richard Bernstein, chairman of the task force, called for the establishment of a pilot program to test the concept.

3c. Clause:

Ms. Towne's secretary, who has been with the firm for 12 years, explained the new procedures to the rest of the staff.

The dealer who offers the lowest price may not be the best choice.

Note: In 3c, no commas around *who offers the lowest price* because this is essential information to identify the dealer referred to.

Comma Rule 5. Use commas to separate words in a series. Use the commas *between* items and not before or after the series. Separate the parts of dates and addresses in sentences.

The assortment of tubes in the exhibit included a diode, a triode, and a pentode.

On June 15, 1995, she moved to 640 S. Montpelier Avenue, Charleston Park, IL 60121.

Sharon, Steve, and I signed up for the seminar in September.

Note: As the third example suggests, use a comma also before the *and* joining the final item to a series. Without this comma, it's not clear whether all three signed up or just "Steve and I."

Semicolon (;)

Use a semicolon between independent clauses in a sentence when no conjunction is present. Clauses are "independent" when they can stand alone. The semicolon is therefore most like a period. The semicolon links related ideas into one sentence; the period divides ideas into separate sentences. Two common patterns follow:

> Peripherals' share of the total installed value peaked between 1975 and 1980; terminals became correspondingly more important in subsequent years.
>
> Software costs exceeded $180 billion in 1993; however, this may be an understated figure because it refers mostly to general purpose computers.

Colon (:)

Use a colon after a general statement to point to specific details. A colon functions like a directional sign pointing to what follows and is often signaled by words such as *thus, as follows,* and *the following.* Don't use a colon if the specific details are not preceded by a complete statement.

> Negotiations touched on the following topics: wages, benefits, safety standards, and leaves of absence.
>
> Negotiations touched on wages, benefits, safety standards, and leaves of absence.

Note: Other uses of a colon include separating the hour and the minutes (8:53 P.M.) and following the salutation in a business letter (Dear Mr. Unger:).

Underlining

Underlining is used in handwritten, typed, or word-processed papers to substitute for the italic type used in printed works. Thus, to italicize, <u>underline</u> the titles of books and the names of magazines, newspapers, movies, TV series, and plays. Also, underline (1) foreign words used in English and (2) words, letters, and figures referred to as such.

> Today's <u>New York Times</u> referred to articles in <u>Omni</u> and <u>Byte</u> and carried a review of Tupper's <u>Essentials of Business Telecommunication.</u>
>
> After shopping at Honest Bob's, I learned the meaning of <u>caveat emptor,</u> which means "let the buyer beware."

The most commonly used English word is the; the most frequently occurring letters are e and i; the most frequently confused numbers are 7 and 9.

Note: Underlining words for emphasis is a matter of style and taste. Too many underlines dilute the emphasis.

Quotation Marks (" ")

Use quotation marks to enclose someone's exact words but not restatements of others' words. Start quotations of complete sentences with a capital letter. Place commas and periods inside the final quotation marks. Four common patterns for presenting quoted material follow.

The chairman said, "Let the meeting come to order."

"Let the meeting come to order," said the chairman, "so we can get to all these important issues."

"Let the meeting come to order," said the chairman.

The chairman said the meeting should come to order. (Not a direct quote.)

Quotation marks are also used to enclose titles of shorter works such as book chapters, stories, articles, essays, reports, songs, and poems.

A reprint of "Electric Drills" from *Consumer Reports* is cited in the chapter entitled "Analytical Reports" in *How to Write for the World of Work.*

Hyphen (-)

Use a hyphen in closely associated words and in compound numbers and to divide words at the end of a line.

Words:

emitter-coupled logic, high-gain amplifier

Numbers:

twenty-one, ninety-nine

Note: In technical writing, you would ordinarily use numerals for numbers greater than nine.

Word Division

Divide between syllables, close to the middle of words, between doubled consonants (plot-ter), after doubled consonants that end a word to which an ending is added (cross-ing), and between two consonants that come between two vowels (ran-dom).

Note: In word-processed or typewritten material today, words are generally not divided at the end of a line but carried over to the next line.

Dash (—)

Use the dash to mark sudden changes of thought that cause a change in sentence structure.

> The chairman—the board had just informed him of his selection—came into the room smiling nervously.
>
> The paint on the front panel of the console—there were four panels in all—showed evidence of heat damage.

Other uses of the dash—and there are several—are largely matters of stylistic choice. In these cases, other punctuation (such as parentheses) can be used.

Note: Most typewriter or word processor keyboards do not offer a dash key. To use the dash, type two hyphens with no space between (--).

Apostrophe (')

Use an apostrophe to show possession and to indicate omitted letters in contractions.

Possession:

> the system's performance, the students' reports

Contraction:

> don't, weren't, can't, it's, you're

Equally important in using the apostrophe properly is *not* to insert it where it doesn't belong. In particular, don't use the apostrophe before the final *s* in plural terms and don't use it in possessive pronouns.

> *Wrong:* the Finchley's arrived, it's completion, the 1990's, our's
>
> *Right:* the Finchleys arrived, its completion, the 1990s, ours

Parentheses [( )]

Set off minor digressions, explanations, or amplifying material in parentheses. Use parentheses, rather than dashes, when the explanatory material does not greatly change the structure of the sentence.

> The chairman (or chairperson, if you prefer) came into the room grinning ear to ear.
>
> The paint on the front panel of the console (a Bassinger 350) showed evidence of heat damage.

In reports, parentheses are also commonly used to enclose numbers or letters of reference and to direct attention to visuals.

> Parentheses may be used for (1) minor digressions, (2) explanations, or (3) amplifying material, such as (see Figure 2).

Brackets ([ ])

Use brackets to enclose your own comments or explanations within someone else's words.

> "Before the development of the modem [an acronym for *mo*dulator-*dem*odulator], systems like these were not possible," Dr. Lopez explained.

When you quote material that contains an error, the Latin word *sic* ("thus") in brackets indicates that the error was in your source:

> According to the report, "The currency should decline an additional $20% [*sic*] against the dollar."

Brackets may also be used to nest material that is already within parentheses or to enclose parentheses themselves, as in mathematical equations.

> The point had been made earlier by C. P. Snow (see *The Two Cultures and the Scientific Revolution* [Cambridge: Cambridge University Press, 1961] for a full discussion).
>
> $3[(a + b)^2 + c] = y$

Slash (/)

Use the slash to indicate options: *and/or, pass/fail, on/off setting, producer/director.* In typing fractions, the slash can separate the numerator and denominator: $3/4$, $2\,2/3$.

Unit 6.4 Exercise

A. In each of the following sentences, insert any necessary *periods, question marks, exclamation points,* and *colons.*

1. What is the source of your information
2. I'd like to know how she reached her conclusions
3. Help.
4. At 930 AM, the personnel department issued the following announcement The Hibbing facility will be closed at 730 PM Wednesday for the Thanksgiving holiday and will reopen at 7 AM Monday
5. They listed three major considerations materials availability, labor costs, and proximity to transportation

6. "Eureka" he shouted "I have it"
7. The declines were limited to two fall months October and November
8. Will the banks be closed this Thursday
9. I wonder if there's time to make a few changes
10. Prof Louis Laskey of MIT received a research grant from Mammoth Data Inc

B. In each sentence, insert commas and semicolons as needed.

1. Really it's all right.
2. After the introductory remarks and the rather drawn out background summary the presentation flowed quite well.
3. Although I voted for the proposal I had my doubts we could finish by July 3 1998.
4. Susan on the other hand thought it could be done by April.
5. The last session that I attended turned out to be the most interesting.
6. I had done some research in this area however I was hardly prepared for a detailed discussion of the plan.
7. Spencer's original suggestion which had won few supporters began to look better and better as the discussion dragged on.
8. The turbulence was fierce during the return flight I found it difficult to finish the paperwork.
9. They served us a tasty little breakfast of juice coffee toast scrambled eggs and bacon.
10. When I got back to the office I finished reading the report you were right about the regulatory issue.

C. In each sentence following, add *quotation marks* as needed and *underline* any terms that should be in italics.

1. That's a fine example of symbiosis in science, said Professor Cavendish. Without Bohr's explication, he continued, there could have been no fission demonstration in 1938.
2. She also assigned Chapter 23, Business, Government, and the Legal System, in Boone and Kurtz's Contemporary Business.
3. The letter q on my old typewriter was broken, so I had to stay away from words like quotient and quantity.
4. What does the it in It is raining refer to?
5. My report The Future of Technology is going to be longer than Tolstoy's War and Peace.

D. Insert *apostrophes* where needed.

1. Its time to call Helens stockbroker again.
2. This markets low point probably hasnt been reached yet.
3. The Wilsons oldest daughters roommates staying with them during the holidays.
4. These word problems solutions are given in the instructors manual, which Prof. Taylors always consulting.

5. Its obvious theirs is the first project of its kind hes seen.

E. Insert *hyphens* (-), *dashes* (—), and *slashes* (/) where needed.

1. My neighbors they're a retired couple in their 70s have become biking enthusiasts.
2. They park their bikes and I don't mean bicycles in the garage and leave their old four door sedan by the curb.
3. As they see it, it's an either or choice, and they've made it.
4. Ninety nine times out of a hundred, they'll suit up for a two block ride to the drugstore rather than walk.
5. Lately they've been talking about skydiving there's a club at the regional airport here and I wouldn't be surprised if my devil may care neighbors start storing parachutes in that garage of theirs.

F. Insert *parentheses* and *brackets* where needed.

1. She cited two objections: 1 the cost and 2 the lack of compatibility with our present equipment.
2. He argued that "the Harlanger Effect named for physicist Werner Harlanger, 1886–1963 is almost certainly a contributing factor to the failure of this design."
3. When we look at a Mercator projection of the region see Figure 4, the disparity in size is evident.
4. The tone of the public response was summarized in a letter received by the board: "You and all pinko *sic* like you should go soak your heads in a pickle barrel."
5. Several researchers have argued for the broader view see Charles McDougall, *The Uncertainty Principle in Subatomic Particle Physics* Palo Alto, CA: Stanford University Press, 1948, 137–42 for a representative example.

UNIT 6.5 SPELLING

Because the English language evolved through so many foreign inputs and because pronunciation has continued to change long after the written form of the words had been fixed, many English words are not spelled as they are pronounced. That's one of the problems we all face as writers.

Another problem is that English spelling *rules* are notoriously inconsistent, complicated,subject to exceptions. If you try to use "*i* before *e,* except after *c,* or when sounded like *a,* as in *neighbor* or *weigh,*" you'll still have to make exceptions for *forfeit, height, leisure, seize, weird, efficient,* and *conscience* . . . to name only a few.

On the positive side of the spelling problem is the fact that most words *are* written as they are pronounced. Therefore, listening to the sound of a word is a good way to determine its spelling in most cases. We should be able to spell *library, nuclear,* and *February,* for example, if we don't mispronounce them as "liberry," "nukelar," and "Febrary" or "Febuary."

For words whose spelling is not expressed in pronunciation, concentrate on the difficult spots. Look for unpronounced letters such as the silent *e* (management), doubled or undoubled letters (perso*nnel,* co*mmittee,* di*lemm*a), and variant spellings of the same sound (li*c*ense, con*s*ensus).

The best way to learn to spell words that cannot be logically analyzed by sound is to keep a list of those that give you trouble. Include the common words from your academic and career field. As you write these words down, concentrate on the special features that make them hard to spell. Notice the silent letters, doubled or undoubled letters, *-ance* or *-ence* endings, and similar features. Then cover the word up and rewrite it. You can also ask a friend or classmate to read words to you while you practice writing them.

When you are asked to write reports, memos, or business letters, you should make sure—with the aid of a dictionary, and a spell checker if available—that all your words are spelled correctly. Beyond that, a mastery of the common troublesome words will help you in cases such as job interviews and in-class writing assignments where use of reference tools is not possible.

The lists of commonly misspelled words that follow in Figures 6–4, 6–5, 6–6, and 6–7 are examples of lists you should compile for yourself. By learning to spell a relatively short list of the most troublesome words, you can reduce your error rate

accommodate
accuracy
achievement
acquire
affiliated
aggressive
allotted
aluminum
analyze
antenna
apparent
appearance
argument
athletics
bankruptcy
believable
boundary
channel
circuit
committee
comparable
competitor
computer
consensus
controlled
definitely
dependent
desirable
development
diaphragm
dilemma
disappear
disappoint
embarrassing
existence
February
forfeit
government
grammar
harass
height
hygiene
indispensable
infinite
innovate
invisible
judgment
labeling
laboratory
library
license
maintenance
management
mathematics
measurement
miniature
miscellaneous
movement
necessary
nickel
noticeable
occasionally
occurrence
omitted
parallel
pastime
penicillin
perceive
permissible
persistence
personnel
possession
precedent
preparation
probably
procedure
proceed
professor
questionnaire
receive
recommend
repetition
resistance
rhythm
schedule
secretary
seize
separate
similar
studying
successful
suppress
surprise
symmetry
technician
temperature
until
vacuum
weird
writing

FIGURE 6–4
General list of words commonly misspelled.

acoustic
alternator
ammeter
ampere
amplifier
amplitude
anode
antenna
armature
attenuation
bandpass
bandwidth
battery
bias
bypass
calibration
capacitance
capacitor
characteristic
Chebyshev
chrominance
circumference
coaxial
Colpitts
convergence
coulomb
crystal
decibel
degenerative
discrete
dissipation
Doppler
efficiency
emitter
enhancement
equation
exponential
filament
fluorescent
focused
frequency
Gaussian
hexadecimal
hybrid
impedance
inductance
intermediate
intermittent
ionization
joule
Kirchhoff
limiter
linear
Lissajous
logarithm
luminance
metallic
mho
microwave
mil
milliwatt
mutual
negligible
neutralize
nucleus
ohmmeter
oscilloscope
permeability
photodiode
piezoelectric
potentiometer
quartz
reactance
rectifier
resonance
rheostat
schematic
series
sideband
signal
silicon
sine
sinusoidal
solder
solenoid
subassembly
superheterodyne
switch
synchronous
thermistor
tolerance
transceiver
transformer
transient
transistor
transmitter
trigonometric
voltage
watt
zener

FIGURE 6–5
Electronics spelling list.

abacus
access
accumulator
addressability
algorithm
allocate
alphanumeric
annotation
array
assembler
asterisk
asynchronous
auxiliary
backup
batch
binary
Boolean
byte
cache
carriage
cartridge
cassette
collate
comparator
compatibility
compiler
concordance
concurrent
continuous
cursor
cybernetics
debug
diagnostic
disk
diskette
dispatching
encryption
executable
expedite
facsimile
firmware
flexible
flowchart
heuristic
hexadecimal
hierarchical
hierarchy
Hollerith
holography
hysterisis
initialization
inputting
inquiry
integer
interleaving
interpreter
interrupt
linkage
mainframe
mantissa
matrix
microfiche
microprocessor
mnemonic
nanosecond
on-line
operand
pagination
parabola
parameter
patching
peripheral
polyphase
precedence
printout
programmable
programmer
programming
pseudocode
quadrant
queue
referencing
repertoire
retrieve
reusable
satellite
sequential
servomechanism
stochastic
synonym
syntactic
syntax
telecommunications
throughput
topology
transparent
turnkey
unitization
Venn
versatile

FIGURE 6–6
Computer spelling list.

accelerated
accrual
allocation
amortization
annuity
appraisal
asset
bailment
bankruptcy
beneficiary
bookkeeping
breach
brokerage
business
carrier
chattel
commercial
commodities
conglomerate
controller
convenience
convertible
cooperative
copyright
correlation
correspondent
cumulative
debenture
deductible
deferral
deficiency
depositor
depreciation
differential
encumbrance
entrepreneurship
equilibrium
equity
expenditure
extraordinary
feasibility
fiscal
franchisee
goodwill
grievance
hazardous
impasse
indemnify
injunction
installment
insurable
intangible
journalize
lease
ledger
leverage
lien
liquidity
marketability
markup
materiality
mediation
memorandum
mortgage
multinational
negotiable
no-load
obsolescence
oligopoly
plaintiff
planning
preemptive
preferred
prepaid
principal
promissory
proprietorship
proxy
punitive
receipt
recession
reciprocity
recruiting
recycling
requisition
rescission
residual
solvency
specialty
statutory
stockholders
subsidiary
tariff
trademark
treasury
turnover
underwriting
variable
variance
warranty

FIGURE 6–7
Business spelling list.

noticeably. Remember, you already use a great many words that you spell with no trouble.

UNIT 6.6 ABBREVIATIONS AND SYMBOLS

Abbreviations and symbols are used to save time, space, and effort for both writers and readers. In technical and business situations, frequently used terms and concepts are often replaced by these shorthand forms. Some abbreviations or symbols are used so often that they become more recognizable than the original terms: % for *percent,* FM for *frequency modulation,* c.o.d. for *collect on delivery,* and VCR for *videocassette recorder,* among others.

But not all abbreviations and symbols are recognizable by everyone. A heavy dose of shorthand forms within a text can leave readers with the impression that they're reading a foreign language. The key to using abbreviated forms effectively is to consider the audience.

Some Guidelines for Using Abbreviations and Symbols

1. In writing for a general audience, use abbreviations and symbols sparingly.

	UNIT	SYMBOL	EQUIVALENT
Frequency	hertz	Hz	cycles per second
Force	newton	N	kilogram-meters per second squared
Pressure	pascal	Pa	newtons per square meter
Energy	joule	J	kilogram-meter
Power	watt	W	joules per second
Quantity of electricity	coulomb	C	ampere-second
Electric potential	volt	V	watts per ampere
Capacitance	farad	F	coulombs per volt
Electrical resistance	ohm	Ω	volts per ampere

FIGURE 6–8
Common technical abbreviations and symbols.

2. In technical and business writing, use the commonly accepted abbreviations and symbols for a field or an industry, as established in journals, textbooks, and company manuals. See Figure 6–8 for some examples.

3. Use fewer abbreviations and symbols in the text of your writing and more in tables, figures, and equations.

4. In business letters and technical or business reports, common abbreviations and symbols in combination with *figures* are appropriate: 45.4%, $15.75 ea, 3,800 rpm, 72 cc, and so on.

5. Most abbreviations for amounts and measures are written lowercase. Exceptions (such as F, C, and V) result from capitalized originals (Fahrenheit, Celsius, Volta).

6. Abbreviations for plural amounts or measures do not add an *s:* 25 bbl (not bbls), 5 oz, 105 yd, 12 min.

7. It is becoming established practice in business and technical writing to omit the period after an abbreviation. This practice simplifies matters, but it should be used with caution in cases that would introduce possible confusion. For example, don't abbreviate *inches, number,* and *ante meridiem* as *in, no,* or *am,* when these might be read as words. Use *in., no.,* and *a.m.* instead. The math abbreviations *tan, cot,* and *sin,* on the other hand, may be appropriate without periods because they would ordinarily occur in equations.

8. Many agencies and public organizations also abbreviate their names in capitals without periods: CIA, IRS, GOP, NASA, UNICEF. The last two are examples of abbreviations that are commonly pronounced as words (nasă, yoonisef). In form, they are acronyms.

A Word About Acronyms (AWAA)

An acronym is a special type of abbreviation composed of the first letters of the several words of a term. These letters then form a "word" or at least a combination that can be pronounced as a word.

Some notable examples are BASIC (*B*eginner's *A*ll-purpose *S*ymbolic *I*nstruction *C*ode), laser (*l*ight *a*mplification by *s*timulated *e*mission of *r*adiation), radar (*ra*dio *d*etection *a*nd *r*anging), and snafu (*s*ituation *n*ormal—*a*ll *f*ouled *u*p). *Laser, radar,* and *snafu* have become so widely used that they have lost their initial all-capitals form and become words in the ordinary sense.

Whether a pronounceable "word" is formed or not, the tendency to abbreviate a sequence of words to a sequence of first letters is very strong in technical and business fields because of the increased speed and ease of reference. As a writer who considers the audience (and realizes that it may include both specialists and nonspecialists), you should follow the practice of defining all acronyms and buzzwords at the point of initial use. Give the original term first, followed by the abbreviation/acronym in parentheses, or vice versa:

> The Mechanized Income Tax System (MITS) does a more efficient job of planning taxes than . . .
>
> The parallel-to-serial conversion of the board was achieved by using the CDC1856 UART (Universal Asynchronous Receiver-Transmitter).

UNIT 6.7 NUMBERS AND EQUATIONS

In technical and business writing, numbers and equations occur more frequently because of the need for precision and for reproducible results. The greater frequency of quantitative data creates a need for effective ways to incorporate numerical forms into verbal contexts.

Handling Numbers

The most common decision is whether to write a number as a figure or as a word. Guidelines for this and other common situations follow.

1. Write numbers 10 and above as figures and the numbers one through nine as words, with the following exceptions.

2. In a series of numbers, some of which are above nine and others below, write all the numbers as figures:

> We took receipt of 4 micros, 65 instruction manuals, and 2 overdue bills.

3. Generally, don't write numbers of a million or more as figures. For large round numbers, use words: *three million, twelve billion,* and so on. For large mixed numbers, use a figure-plus-word combination: *12.8 billion, 23 million.*

4. Numbers that start a sentence are always written as words: "Three hundred twenty-one calls came in," *not* "321 calls came in."

5. Use a hyphen (-) in all compound numbers from *twenty-one* to *ninety-nine* when you must write them as words. Use a hyphen in fractions used either as nouns or modifiers.

three-fourths of the space	a *three-fourths* majority
a proportion of *two-fifths*	a *two-fifths* proportion

6. Use figures for common *applications* of numbers to units of measurement, prices, times of day, dates, decimal amounts, percentages, temperatures, statistics in tables, and other data. Also use figures with units of time such as minutes, hours, days, and weeks:

2 inches by 4 yards	23 June 1997
$128.75	4.143
8:35 P.M.	26.5°C
74.8%	

7. Use decimals rather than fractions if the decimal allows greater precision: 3.34 rather than $3\frac{1}{3}$. Use fractions rather than decimals if the fraction is more precise: $8\frac{13}{15}$ rather than 8.867. For decimals with a value less than one, use a zero before the decimal point to avoid confusion with preceding material: (0.18).

8. Use figures and words alternatively to distinguish adjacent numbers from each other: twelve 10-inch shelves, 8 two-liter bottles, three 400-meter sprints.

Handling Equations

If your report includes only a few simple equations or formulas, incorporate them into the text using a colon or parenthesis:

> Payment was calculated (17 × $15.50 = $263.50) according to the formula given in the sales manual: units × adjusted price = prepaid amount.

If your report includes a series of equations that are relatively complex or that deserve special attention, set them off from the text:

1. Center them across the page.
2. Number them consecutively in parenthesis at the right of the page.

3. Leave a space above and below each equation you set off.
4. Identify the symbols used in the equation.
5. Use the same punctuation (or no punctuation) no introduce the equation as you would if the equation were not set off (see the following examples).

The equation for a straight line is

$$Y = a + bX \tag{3}$$

where Y = dependent variable
a = Y-intercept
b = slope of line
X = independent variable

For this summing amplifier circuit, the output voltage is derived as follows:

$$V_o = -\left(\frac{R_F}{R_1} V_1 + \frac{R_F}{R_2} V_2\right) \tag{7}$$

where V_o = output voltage
R_F = feedback resistance
V_1 = input voltage 1
V_2 = input voltage 2
R_1 = input series resistance for V_1
R_2 = input series resistance for V_2

Appendix: Answers to Exercises

INTRODUCTION: TECHNICAL STYLE

Note: Answers to this exercise are models; other good solutions are possible.

1. a. Our new fax system is faster, cheaper, and more reliable than express mail or couriers.

b. The company's restrictive personnel policies seemed to stifle initiative and creativity.

c. When I first met Dr. Chesterton, he said we could improve the survey design.

d. Because of gasoline price increases, we considered buying a fleet of electric cars for residential deliveries.

e. Despite the electronic monitors, our downtown cash machines will probably continue to experience tampering until we dramatically increase security precautions.

2. a. The Bears offered Flanders $950,000 over two years; Flanders wanted $1.6 million and a three-year contract.

b. The Xtraordinary-8 laser printer gave us higher resolution at a higher speed and a lower price than the old Ordinary model.

c. We're down to our last carton of sterile gauze and a bare dozen tongue depressors. We need to restock all high-use items by the time the clinic opens Friday.

d. The collision of a cold front with warm, moist air from the Gulf spawned a series of tornadoes in three southern states.

e. After three quarters of slow sales, the company's restyled men's clothing line generated record profits of $3.75 million during the holiday season.

3. Permanent magnets, which are made of hard magnetic materials, are magnetized by induction during manufacture. When the magnetizing field is removed, residual induction keeps the materials magnetized indefinitely. Of course, magnets subjected to high temperature may lose their properties. At 800°C, for example, iron gives up its "permanent" magnetized state.

4. **a.** Language would likely refect efforts to define technical terms as they are introduced and to orient the reader to the technical context.
   b. The report might give up its highly structured and segmented format for a more general development of the topic according to the purpose.
   c. The lab report presumably was written to provide a precise and objective description of the experimental situation. The revised report might include less descriptive detail on the equipment and the experimental procedure. More emphasis might be devoted to analysis of results and to their significance in the technical context.

6.1 BASIC GRAMMAR

A.

1. F	**6.** S
2. R	**7.** S
3. S	**8.** F
4. R	**9.** R
5. F	**10.** F

B.
1. Ordering 6 weeks before the due date, Thelma received the parts in plenty of time.
2. A new sedan with front-wheel drive was unveiled at the fall auto show.
3. Hoping he was right, Randy diagnosed the problem as a twisted cable.
4. Sweeping the field with binoculars, Millicent noticed a slight movement in the hedgerow.
5. In the foundation, the inspector found hairline cracks that hadn't been there yesterday.
6. Refusing to concede any of their advantages, the two sides kept the problem deadlocked between them.
7. Careful positioning of the ceiling mirrors made the back of the store visible from the front counter.

C.
1. Check the oil, rotate the tires, and replace the air filter.
2. We must either hire a new secretary or ask Gary to work overtime for two weeks.
3. This package is not only more powerful but also easier to use.
4. I stayed with my cousin, who is an electrical engineer and a gourmet cook.
5. I liked the movie for its characters, plot, and outstanding cinematography.
6. The more I practiced this technique, the more I enjoyed skiing.
7. It may be more productive to go back to the beginning than try to find the way out from here.

D. **1.** were
2. have
3. is
4. concerns
5. requires
6. has
7. was
8. remains
9. are
10. make

E. **1.** *last* his
2. *Deborah* her
class it
3. *Bill* he
problem it
4. *Grigsby* me
neither she
5. *Petrakis* his
6. *Harriet* her
7. *Mr. Philby* whom
8. *contracts* which
9. *She* whom
10. (*unspecified antecedent*)
whoever

F. **1.** *she* The student admitted her mistake to the professor.
2. *this* Jessica was bored by the long discussion on computer modeling.
3. *it* The *Times* says good tickets for the concert are still available.
4. *it* The systems department office, which is being remodeled today, is located in the annex building.
5. *which* The surge in consumption overloaded the system and led to the power outage.
6. *they* Since the city council passed the snow ordinance, cars on our street have been towed every time a few snowflakes fall.
7. *it* After unbolting the metal plate with a wrench from his tool kit, he put the plate on the counter (or: . . . put the wrench/the kit . . .).

G. **1.** benches
2. stereos
3. passersby
4. feet
5. companies
6. analyses
7. displays
8. leaves
9. criteria
10. series

H. **1.** attorneys
2. ours, company's
3. whose, everyone's
4. its
5. band's, grounds
6. Hartleys', friends
7. shipment's, their, truckers'
8. members', your
9. Wilson's, hers
10. its, media

6.2 VOCABULARY AND USAGE

A. **1.** number, intramural
2. well, good
3. rise, affect
4. May, continual
5. percentage, weather
6. lie, lain
7. fourth, than
8. principal, its
9. among, capitol
10. fewer, likely

11. farther, ensure
12. sit, you're, hear
13. its, brakes
14. accept, device
15. very, complements
16. effect, morale
17. implied, except, personnel
18. risen, passed
19. may be, they're
20. proceed, criteria, ideal

B.
1. I hope there is a good turnout for the meeting tomorrow.
2. He was so annoyed by the questioning, he stalked off the platform.
3. Because of the early frosts, produce will be more expensive this winter.
4. They are very disappointed about this antitrust case.
5. I suppose you will want to see the written report before you decide.
6. Give your work experience in reverse chronological order (i.e., in reverse order of occurrence).
7. A lot of the students left early for the holidays.
8. Joan's oral presentation, which she had practiced delivering for several days, was very well received.
9. The room was so large, I could hardly see the diagrams he was drawing on the board.
10. We should try to make a damage assessment before the press conference.
11. Bring me the manual that gives the assembly instructions for this model.
12. The position offers several unusual benefits (e.g., a polo club membership and a free subscription to *Sport of Kings* magazine).
13. The power went out because a fire broke out in the switching station in Decatur.
14. Why don't we use the boardroom for the meeting on Friday?
15. *The Case for Artificial Intelligence,* by Przybski et al., was placed on reserve in the library.

6.3 CAPITALIZATION

1. After he read the article in *Consumer Reports,* Sam visited three dealers to price a Chrysler minivan.
2. She prepared her report on a Compaq ProLinia personal computer.
3. The subway entrance was at the corner of State and Madison streets.
4. In the first debate, the Democratic candidate was chosen by lot to make the opening statement to the audience in Burton Auditorium.
5. Weakening markets in Europe and South America were beginning to affect the Japanese economy.
6. My appointment is in the Wrigley Building, which is that graceful white building across the river.
7. Now that she was a bank vice president, she began to sympathize with Vice President Nelson of the First National Bank, who was always being pursued for investment advice by his neighbors.
8. At the reception, I ran into Dr. Washburn, my dentist.
9. The invitation list had failed to include the president of the Cambridge Supply Company, our new subcontractor.

10. Norman decided to pursue a master's in electrical engineering instead.
11. Later she enrolled in the M.B.A. program at the same university where she had earned her bachelor's.
12. This project is due at the end of the spring semester.
13. This year, Independence Day falls on a Monday, so we'll have a three-day weekend for the trip to Grandma's.
14. To get to the park entrance, you have to go northwest from here for about five miles.
15. This will be her first trip to the West Coast since June of last year.
16. My adviser said that if I want to major in business, I'll need to take more courses in economics, accounting, science, algebra, and Spanish.
17. The new class schedule I worked out includes Cost Accounting II, College Algebra, and Data Base Management Systems II.
18. We're going to meet with two of the analysts in the Marketing Research Department to discuss the survey results.
19. The press secretary issued the following statement: The administration believes that the article in *The New York Times* was premature and based on erroneous information.
20. Professor Carter assigned Chapters 2 and 3 in addition to the Stoner article, "A Somatic Mutation Theory," in the *Journal of Psychosomatic Research.*
21. I told the lieutenant that Captain Abrahams wanted to see him.
22. She brought along her uncle, Walter Mitty, to meet Aunt Alice.
23. She kept saying, "Oh, Mother," all through dinner.
24. They planned a canoe trip on the chain of lakes at the Canadian border.
25. Next Saturday is the anniversary of the Mount Saint Helens eruption.

6.4 PUNCTUATION

A.
1. information?
2. conclusions.
3. Help!
4. 9:30 A.M.
   announcement:
   7:30 P.M.
   7 A.M. Monday.
5. considerations:
   transportation.
6. "Eureka!"
   shouted.
   it!"
7. months:
   November.
8. Thursday?
9. changes.
10. Prof.
    Inc.

B.
1. Really,
2. summary,
3. proposal, July 3,
4. Susan, hand,
5. (no change)
6. area; however,
7. suggestion, supporters,
8. flight;
9. juice, coffee, toast, eggs,
10. office, report;

C. **1.** "That's . . . science,"
"Without . . . explication,"
"there . . . 1938."
2. "Business . . . System,"
<u>Contemporary Business.</u>
3. <u>q</u>; <u>quotient</u>; <u>quantity</u>.
4. it, "It is raining"
5. "The Technology", *War and Peace*

D. **1.** It's, Helen's
2. market's, hasn't
3. Wilsons', daughter's, roommate's
4. problems', instructor's, Taylor's
5. It's, he's

E. **1.** neighbors—they're, 70s—have
2. bikes—and, bicycles—in, four-door
3. either/or
4. Ninety-nine, two-block
5. skydiving—there's, here—and, devil–may–care

F. **1.** (1)
(2)
2. [named for physicist Werner Harlanger, 1886–1963]
3. (see Figure 4),
4. [*sic*]
5. (see . . . [Palo Alto . . . 137–42] . . . example).

Index